AF869758

Design Science and Innovation

Series Editor

Amaresh Chakrabarti, Centre for Product Design and Manufacturing, Indian Institute of Science, Bangalore, India

The book series is intended to provide a platform for disseminating knowledge in all areas of design science and innovation, and is intended for all stakeholders in design and innovation, e.g. educators, researchers, practitioners, policy makers and students of design and innovation. With leading international experts as members of its editorial board, the series aims to disseminate knowledge that combines academic rigour and practical relevance in this area of crucial importance to the society.

More information about this series at https://link.springer.com/bookseries/15399

D. Bijulal · V. Regi Kumar · Suresh Subramoniam ·
Rauf Iqbal · Vivek Khanzode
Editors

Technology-Enabled Work-System Design

Select Proceedings of HWWE 2018

Springer

Editors
D. Bijulal
Department of Mechanical Engineering
Government Engineering College Barton Hill
Trivandrum, Kerala, India

Suresh Subramoniam
School of Management
College of Engineering Trivandrum
Trivandrum, Kerala, India

Vivek Khanzode
National Institute of Industrial Engineering
Mumbai, Maharashtra, India

V. Regi Kumar
Department of Mechanical Engineering
College of Engineering Trivandrum
Trivandrum, Kerala, India

Rauf Iqbal
National Institute of Industrial Engineering
Mumbai, Maharashtra, India

ISSN 2509-5986 ISSN 2509-5994 (electronic)
Design Science and Innovation
ISBN 978-981-16-1883-3 ISBN 978-981-16-1884-0 (eBook)
https://doi.org/10.1007/978-981-16-1884-0

© The Editor(s) (if applicable) and The Author(s), under exclusive license to Springer Nature Singapore Pte Ltd. 2022, corrected publication 2022
This work is subject to copyright. All rights are solely and exclusively licensed by the Publisher, whether the whole or part of the material is concerned, specifically the rights of translation, reprinting, reuse of illustrations, recitation, broadcasting, reproduction on microfilms or in any other physical way, and transmission or information storage and retrieval, electronic adaptation, computer software, or by similar or dissimilar methodology now known or hereafter developed.
The use of general descriptive names, registered names, trademarks, service marks, etc. in this publication does not imply, even in the absence of a specific statement, that such names are exempt from the relevant protective laws and regulations and therefore free for general use.
The publisher, the authors and the editors are safe to assume that the advice and information in this book are believed to be true and accurate at the date of publication. Neither the publisher nor the authors or the editors give a warranty, expressed or implied, with respect to the material contained herein or for any errors or omissions that may have been made. The publisher remains neutral with regard to jurisdictional claims in published maps and institutional affiliations.

This Springer imprint is published by the registered company Springer Nature Singapore Pte Ltd.
The registered company address is: 152 Beach Road, #21-01/04 Gateway East, Singapore 189721, Singapore

Preface

We have a great pleasure to bring out the Proceedings of the 16th International Conference on Humanizing Work and Work Environment 2018 (HWWE 2018) with the theme "Technology Enabled Workplace Design" held during December 14–16, 2018 at College of Engineering, Trivandrum (CET), Kerala. This was organized jointly by the Department of Mechanical Engineering and CET School of Management, College of Engineering Trivandrum, Kerala. The HWWE series of Conferences is an annual flagship event of the Indian Society of Ergonomics (ISE) held in association with International Ergonomics Association (IEA). The Trivandrum Chapter of the Indian Institution of Industrial Engineering was the knowledge partner for HWWE 2018.

The 16th edition of HWWE aimed to bring together researchers from various countries working in different areas of Human Factors Engineering, to discuss the current developments, opportunities, and challenges in their respective fields. The HWWE 2018 received overwhelming response with nearly 185 submissions for presentation in the Conference and out of which only 145 submissions got selected after reviewing by renowned academicians and practitioners hailing from leading institutions from India and abroad.

The keynote speakers representing the International Ergonomics Association were Prof. Jose Orlando Gomes, Federal University of Rio de Janeiro and the Vice-President and Treasurer of IEA from Brazil, Prof. Eva Honeyman Vice Chairperson, Israel Ergonomics Association and Head of Ergonomics at Ergo group and Prof. Peter Honeyman from Israel. Other keynote speakers include Dr. Debkumar Chakrabarti IIT Guwahati and Vice-President of ISE, Dr. A. K. Ganguli, President, ISE, Prof. P. C. Dhara, Hon. General Secretary, ISE. Various lead lectures were presented by academicians and industry experts during the technical sessions. These lectures were delivered by Prof. Dibakar Sen, IISc. Bangalore, Dr. Anirudha Joshi, IIT Bombay, Mr. Guenter Fuhrmann, M/s. Ergoneers, Germany, Dr. Rauf Iqbal, NITIE, Prof. M. Muzammil, Aligarh Muslim University, Dr. Nandita Bhattacharyya, Assam Agricultural University, Dr. Ajita D. Singh, Punjab University, Dr. S. Mukherjee, University of Calcutta, Dr. Sudesh Gandhi, Haryana Agricultural University, Dr. Sougata Karmakar, IIT Guwahati, and Dr. Abid Ali Khan, Aligarh Muslim University.

The Organizing Secretary for the conference was Dr. D. Bijulal, Associate Professor, Department of Mechanical Engineering, CET and Vice Chairman, IIIE, Trivandrum Chapter. The conference was chaired by Dr. A. Samson, Head, Department of Mechanical Engineering and Dr. Suresh Subramoniam, Director, CET School of Management under the patronage of Dr. Jiji C. V., Principal, College of Engineering Trivandrum.

This Proceedings is the outcome of various contributors including the authors, the expert reviewers, the editors, and the publishers. This Proceedings contains 17 number of and outstanding selected papers from those presented at the conference. These manuscripts have undergone rigorous scrutiny in different stages of selection by experts in the field prior to publication. We hope that these papers will find good readership and will receive deserving attention from the Human Factors and Ergonomics fraternity across the world.

The editors wish to thank Dr. Usha Titus, Principal Secretary Higher Education and Former Vice Chancellor, APJ Abdul Kalam Technological University, Kerala and Dr. K. P. Indira Devi, Director of Technical Education for their constant encouragement. The financial support from Antrix Corporation, the Directorate of Technical Education, Kerala and the Centre for Engineering Research and Development (CERD) under the APJ Abdul Kalam Technological University are also greatly acknowledged. We thank all the authors for contributing and giving consent for their latest work to be published in this Proceedings. Special thanks go to reviewers for their critical review. We express our sincere gratitude to all the editorial staff of Springer for their immense support in bringing out this wonderful compendium of selected papers.

Trivandrum, India — D. Bijulal
Trivandrum, India — V. Regi Kumar
Trivandrum, India — Suresh Subramoniam
Mumbai, India — Rauf Iqbal
Mumbai, India — Vivek Khanzode

The original version of this book was revised: The editors name "Bijulal D." and "Regi Kumar V." has been changed globally as "D. Bijulal" and "V. Regi Kumar". The correction to this book can be found at https://doi.org/10.1007/978-981-16-1884-0_18

Contents

Editors and Contributors

About the Editors

Prof. D. Bijulal is a Professor in the Department of Mechanical Engineering currently working in the Government Engineering College, Barton Hill, Thiruvananthapuram, India. He was earlier associated with the College of Engineering Trivandrum, India, for more than 20 years including three years as Professor in the CET School of Management, out of 22 years of teaching experience. He did his M.Tech. in Industrial Engineering and Management from Indian Institute of Technology (IIT) Kharagpur and completed his Ph.D. in Industrial Engineering and Operations Research from IIT Bombay.

The research interests of Prof. D. Bijulal include Industrial Engineering and Management, General Management, Intellectual Property Right, Behavioural Finance, System Modelling and Simulation, System Dynamics, etc. He is an approved guide under the APJ Abdul Kalam Technological University, Thiruvananthapuram and has five scholar doing their research under his guidance. Apart from the publications that he authored, he is a research contributor as reviewer for many international journals and conferences.

Dr. V. Regi Kumar is a Professor in Mechanical Engineering, College of Engineering, Trivandrum, Kerala, India. Dr. V. Regi Kumar acquired his MTech from University of Kerala and MBA from IGNOU. He has done his Ph.D. in Industrial Engineering and Management from the University of Kerala and having seven years of industrial experience and more than 20 years of teaching experience. His research areas include humanitarian supply chains, ergonomics of heavy vehicle drivers, inventory optimization of perishable edibles, emotional ergonomics, niche banking and green supply chains. He is having more than hundred publications and guiding seven Ph.D. scholars.

Prof. Suresh Subramoniam is the Professor and Director of the School of Management, College of Engineering, Trivandrum affiliated to the A P J Abdul Kalam

Technological University, Kerala, India. He has done his B.Tech. in Mechanical Engineering from Kerala University, M.S. in Industrial Engineering from Louisiana State University in USA specializing in the field of Ergonomics and Human Factors Engineering, and his Ph.D. in Management in the area of New Industrial Engineering from Kerala University. Prof Subramoniam's research focuses on human factors engineering including manual materials handling, design of tools/equipment, usability studies and New Industrial Engineering. He has authored more than a hundred papers in journals and conferences, and has authored one book. He has worked in Indian Space Research Organization and also in Universities in Saudi Arabia and Oman and has produced Six Ph.D.'s. He is a Fellow of Indian Institution of Industrial Engineering.

Rauf Iqbal is a researcher, teacher and consultant in Ergonomics and Human Factors Engineering. He is an Associate Professor and In-Charge of Ergonomics Laboratory at the National Institute of Industrial Engineering (NITIE), Mumbai. He holds a Ph.D. degree in Ergonomics. He teaches Ergonomics and Human Factors, Layout and Facilities Planning and Worksystem design to the master's students of Industrial Engineering, and Manufacturing Management. and his research interest is Worksystem design, Engineering Anthropometry and workspace design, Biomechanics of human movement, Occupational ergonomics and Safety management.

Rauf Iqbal has publications of over 120 as journal articles, edited books, book chapters and conference proceedings. He has been regularly presenting and chairing sessions at International conferences over the last 15 years. He has guided several scholars for the Ph.D. degree as well as master's theses. He has reviewed articles for various International journals and also has examined Ph.D. thesis from national and international universities. He is member of various international and national committees like,—BRICS Plus Executive Committee for Human Factors and Ergonomics, Council member of Asian Council on Ergonomics and Design, Ergonomics sectional committee for Bureau of Indian Standards, Executive committee member of Indian Society of Ergonomics etc. He has carried out various projects funded by the Government as well as International Labour Organization. He has been conducting training and consultancy for numerous industries in the areas of ergonomics, work study, work systems design and Safety management.

Vivek Khanzode is Associate Professor at National Institute of Industrial Engineering, Mumbai. He holds M.Tech. and Ph.D. in Industrial Engineering from IIT Kharagpur. He teaches Operations and Supply Chain Management, Worksystem Design, Simulation Modeling and Lean Manufacturing in postgraduate programs at NITIE. He has designed and launched executive and leadership programs in manufacturing and logistics domain. Prof. Khanzode is associated with manufacturing industries for productivity improvement projects using lean manufacturing, process control and robust optimization techniques. His research interests include modeling, design, and optimization of manufacturing and logistics systems.

Contributors

S. Abhinav Department of Mechanical Engineering, NIT Calicut, Calicut, India

S. Abishek Department of Mechanical Engineering, NIT Calicut, Calicut, India

Saleem Ahmed Vellore Institute of Technology (VIT), Vellore, India

S. A. Basha Department of Mechanical Engineering, Rajiv Gandhi University of Knowledge Technologies, Nuzvid, Andhra Pradesh, India

N. Bhattacharyya Department of Family Resource Management and Consumer Science, College of Community Science, Assam Agricultural University, Jorhat, Assam, India

S. A. Binoosh Department of Mechanical Engineering, College of Engineering Trivandrum, Thiruvananthapuram, Kerala, India

R. Borah Dept. of Family Resource Management and Consumer Science, College of Community Science, Assam Agricultural University, AAU, Jorhat-13, India

K. M. Jishnu Devadas Department of Mechanical Engineering, College of Engineering Trivandrum, Thiruvananthapuram, Kerala, India

Francis J. Emmatty Industrial Engineering Laboratory, Department of Mechanical Engineering, National Institute of Technology Calicut, Calicut, Kerala, India

Z. Shah Fahad Mechanical Engineering Department, Aligarh Muslim University, Aligarh, India

M. Shah Faizan Mechanical Engineering Department, Aligarh Muslim University, Aligarh, India

Nikhil Ghag National Institute of Industrial Engineering, Mumbai, India

Solomon Darius Gnanaraj Vellore Institute of Technology (VIT), Vellore, India

Rauf Iqbal National Institute of Industrial Engineering, Mumbai, India

Nusky Ishrat Aligarh Muslim University, Aligarh, India

Karuna Jain National Institute of Industrial Engineering, Mumbai, India

M. Kalita Dept. of Family Resource Management and Consumer Science, College of Community Science, Assam Agricultural University, AAU, Jorhat-13, India

Mukesh Kumar Kamti National Institute of Industrial Engineering, Mumbai, India

Abid Ali Khan Aligarh Muslim University, Aligarh, India

G. Kiran Vellore Institute of Technology (VIT), Vellore, India

Udakam Vamsi Krishna Department of Mechanical Engineering, Rajiv Gandhi University of Knowledge Technologies, Nuzvid, Andhra Pradesh, India

K. Krishnakumar Department of Mechanical Engineering, CET, Thiruvananthapuram, India

Potturi Phani Kumar Department of Mechanical Engineering, Rajiv Gandhi University of Knowledge Technologies, Nuzvid, Andhra Pradesh, India

Vishnu V. Kumar Department of Mechanical Engineering, NIT Calicut, Calicut, India

Richard Laurence Vellore Institute of Technology (VIT), Vellore, India

S. Logeswari Department of Family Resource Management and Consumer Science, Community Science College & Research Institute, TNAU, Madurai, Tamil Nadu, India

K. Manivannan Vellore Institute of Technology (VIT) Vellore, Vellore, India

Vahid Mohammad Department of Mechanical Engineering, NIT Calicut, Calicut, India

Kiran Mohan Department of Mechanical Engineering, NIT Calicut, Calicut, India

Mohammad Muzammil Mechanical Engineering Department, Aligarh Muslim University, Aligarh, India

Balakrishnan Nair HSE Trainer and Consultant, Former Company Representative, Petroleum Development Oman, Muscat, Oman

B. Nallakurumban Department of Family Resource Management and Consumer Science, Community Science College & Research Institute, TNAU, Madurai, Tamil Nadu, India

S. Nirmal Industrial Engineering Laboratory, Department of Mechanical Engineering, National Institute of Technology Calicut, Calicut, Kerala, India

P. K. Padmanathan Department of Farm Machinery, Agricultural Engineering College, TNAU, Kumulur, Tamil Nadu, India

Vinay V. Panicker Industrial Engineering Laboratory, Department of Mechanical Engineering, National Institute of Technology Calicut, Calicut, Kerala, India

P. Parimalam Department of Family Resource Management and Consumer Science, Community Science College & Research Institute, TNAU, Madurai, Tamil Nadu, India

V. Madhusudanan Pillai Department of Mechanical Engineering, NIT Calicut, Calicut, India

Ashok K. Pundir National Institute of Industrial Engineering Mumbai, Mumbai, India

Asif Qureshi KIT College of Engineering Kolhapur, Kolhapur, India; Vellore Institute of Technology (VIT) Vellore, Vellore, India

K. Venkata Rahul Industrial Engineering Laboratory, Department of Mechanical Engineering, National Institute of Technology Calicut, Calicut, Kerala, India

O. R. Rohith Raj Department of Mechanical Engineering, NIT Calicut, Calicut, India

Bhawana Rathore National Institute of Industrial Engineering Mumbai, Mumbai, India

Pubali Saikia Department of Family Resource Management and Consumer Science, College of Community Science, Assam Agricultural University, Jorhat, Assam, India

M. Satheeshkumar Department of Mechanical Engineering, CET, Thiruvananthapuram, India

K. Shashank Siddaveeraiah Industrial Engineering Laboratory, Department of Mechanical Engineering, National Institute of Technology Calicut, Calicut, Kerala, India

Buddana Divya Sri Department of Mechanical Engineering, Rajiv Gandhi University of Knowledge Technologies, Nuzvid, Andhra Pradesh, India

Salma Sultana Industrial Engineering Laboratory, Department of Mechanical Engineering, National Institute of Technology Calicut, Calicut, Kerala, India

Sumaiya Mechanical Engineering Department, Aligarh Muslim University, Aligarh, India

Amal Sunny Department of Mechanical Engineering, NIT Calicut, Calicut, India

Priyanka Thalor Industrial Engineering Laboratory, Department of Mechanical Engineering, National Institute of Technology Calicut, Calicut, Kerala, India

A. Ajay Vishnu Industrial Engineering Laboratory, Department of Mechanical Engineering, National Institute of Technology Calicut, Calicut, Kerala, India

Chapter 1
A Changing World: Adapting to an Ageing Population in the Workplace

N. Bhattacharyya and Pubali Saikia

The world's population is ageing: most countries around the world are in the midst of demographic ageing; virtually every country in the world is experiencing growth in the number and proportion of older persons in their population (Kinsella and Gist 1998). Projected increase in both the absolute and relative size of the elderly population in many third world countries is a subject of growing concern for public policy (Kinsella and Velkoff 2001; World Bank 2001; United Nations 2002; Bordia and Bhardwaj 2003; Liebig and Irudaya 2003). By 2050, two billion people will be aged 60 or over, a proportion that brings with it many implications for society. The Organisation for Economic Co-operation and Development (OECD) estimates that over the next 50 years, its member countries will see a steep increase in the share of elderly persons in the population, as well as a steep decline in their prime working-age populations.

This has been especially true in case of developing countries like India, where the elderly population is increasing rapidly. In the forthcoming decades, there will be a tremendous increase in the number of elderly in India, with their rate of increase being faster than that of the total population (WHO 1984).

1 Ageing: The Indian Scenario

India, like many other developing countries in the world, is presently witnessing rapid ageing of its population. According to World Population Prospects, UN Revision, 2006, the population of aged in India is currently the second largest in the world. Even though the proportion of India's elderly is small compared with that of developed countries, the absolute number of elderly population is on the high. There has been

N. Bhattacharyya (✉) · P. Saikia
Department of Family Resource Management and Consumer Science, College of Community Science, Assam Agricultural University, Jorhat, Assam, India

© The Author(s), under exclusive license to Springer Nature Singapore Pte Ltd. 2022, corrected publication 2022

D. Bijulal et al. (eds.), *Technology-Enabled Work-System Design*, Design Science and Innovation, https://doi.org/10.1007/978-981-16-1884-0_1

tremendous increase in the number of elderly population since independence in India from 20.19 million in 1951 (5.5% of the total population) to 43.17 million in 1981 and 55 million in 1991. According to 2001 census, around 77 million population is above 60 years, which constitutes 7.5% of the total population of the country. This number is expected to increase to 177.4 million in 2025. (The growth rate of the population (1991–2001) of elderly has been higher (2.89) than overall growth rate (2.02) of the total population. India's older population will increase dramatically over the next four decades. According to World Population Data Sheet 2002, 4% of the Indian population are in the age group of 65 plus, which accounts for 41.9 million. This phenomenon of growing population of senior citizens has been the result of recent successes in the achievement of better health standards and a longer span of life for our citizens (UN 2011; Shakuntala 2013).

The impact of this transition means that there will also be a change in the labour supply. Many industries will have to adjust either by bringing more of the youth population into their workforce or by attracting more workers from the older population segment. People need to be working beyond traditional retirement ages.

2 Work Ability and Ageing Workforce

Globally, an ageing workforce has significant implications for developed as well as developing countries. As people age, they become more likely to acquire a disability or other age-related health condition that may reduce their functional capacity and affect their ability to remain in the workforce (Heidkamp et al. 2012).

Heidkamp et al. (2012) further reported that each individual ages differently, and the effects of ageing on physical changes may be:

- Reduced strength. With an increase in age on average, a person's physical strength decreases about 25% by 65 of age.
- Reduced aerobic capacity. An older person gets tired more easily under exertion than a younger, may be of 20 years of old.
- Reduced ability to handle shift work. Older people have less resistance to the stresses of night work and become more easily tired.
- Reduced ability to regulate body heat through sweat and water retention, making older workers more susceptible to heat stroke.
- Reduced reaction time. Indicates learning and remembering things for an older worker takes more time.
- Chronic health conditions. With an increase in age, chronic health conditions may also increase, affecting a worker's ability to do certain jobs. These include diabetes, irritable bowel syndrome, arthritis, high blood pressure, and obesity.
- Vision impairments. Change in vision power is very common as age increases. These changes may cause decreased ability to focus, see in conditions of low light, adjust to changes in light conditions, and distinguish colours in different regions.
- Hearing impairments. Hearing loss with age is well known.

The coming together of ageing, disability, and employment results in a complex set of issues for both older workers and their employers. Encouraging older workers to remain in the workforce will require strategies to accommodate their changing abilities. To do so interventions, facilitating the active performance of work until retirement age is required. According to Finnish researchers, Tuomi et al. (1998), 'a comprehensive solution for the ageing challenge at the workplace is the promotion of work ability during ageing'. According to them, the concept of work ability comprehensively describes 'the way to achieve a better correspondence between ageing and work, has been created in Finland as of the end of the twentieth century. It also includes an objective measurement of work ability based on the index of work ability'. This concept focuses on both human resources and working conditions. The core dimensions of human resources include health, physical and mental capacities and social functioning, competencies, as well as attitudes and values. The core dimensions of work cover the contents and demands of work, physical, ergonomic and psychosocial work environment, as well as management and leadership issues. Work ability is also connected to the microenvironment outside the workplace (family, relatives, friends, etc.) as well as with the macro-environment (infrastructure, services and other societal dimensions). The new core concept of work ability emphasizes the balance between human resources and work.

3 The Present Study (Tea Industry and the Workers)

The concept of work ability relates to the capacity a worker has to perform his work tasks, given his work demands, health status and physical and mental abilities. 'Work ability' is a complex construct reflecting the individual and occupational factors influencing a person's ability to cope in working life. It reflects the interaction between mental and physical activities and worker's functional capabilities, health and subjective assessment of their status in given organizational and social conditions (Tuomi et al. 1998). Generally, work ability is an individual quality, whereas maintaining work ability deals more with actions aiming at promoting an employee's work ability (Julin et al. 2001). Work ability is largely based on individual qualities of physical, mental and social functional capacities, which all consist of various factors: physical functional capacity includes endurance (aerobic and anaerobic), muscle strength (endurance, speed and strength) and skills of the nervous system (flexibility, balance and agility). The improvement of work ability of workers is believed to be economically beneficial to the work place. Good work ability of workers improves the quality of work, resulting productivity and contributes to a better health-related quality of life. According to Ilmarinen et al. (1998), 'being productive at work and being in good health are considered important determinants of prolonging working life and also of retirement in good health'.

Tea is consumed in about 100 countries, and India is the largest producer and consumer of tea in the world (Baroowah 2006). Assam is the largest tea producing state in India and contributes about 60% of the total production of India (Sen 2008).

The majority of workforce in tea fields are women workers (Bhattacharyya et al. 2013). ‘Women workers are involved at every stages of production i.e., from nursery development to the packaging work’ (Lama 1983). The tea planting industry is labour-intensive and workers being exposed to the vagaries of terrain and climate, in addition to exposure to chemicals in the form of pesticides and fertilizers, the possibility of safety and health hazards is very high. They are subjected to adverse working conditions inherent in the work process, viz., adopting awkward postures reaching for the new shoots to be plucked which involve long hours of standing, reaching or bending, repetitive movements and carrying heavy loads on back during plucking operation, etc. Moreover, workers need to travel long distance by carrying the plucked leaves on their head after completion of the entire shift to unload their daily plucked leaves to the place where plucked leaves are being weighed. The workers perform the works by following the same age-old traditional work practices without much awareness about its impact on health. Due to this backbreaking and drudgery-prone work, the work efficiency of the workers decreases, and health of the workers is also affected. Specifically, it aggravates when ageing workers are concerned because of physical effects of ageing. This category of workers has different types of work-related health problems leading to reduced work ability, early retirement, absenteeism and lowered productivity. As opined by Ilmarinen (2001), ‘the improving work ability is one of the most effective ways to enhance the ability and preventing disability and early retirement’.

Thus, the assessment of work ability was considered as important tool as it would measure the ability of workers to perform their jobs, taking into account the specific psychosocial and physical work-related factors, mental and physical capabilities and health. Keeping it in view, the present study was undertaken in tea fields of Assam to assess the Work Ability Index (WAI) of workers of above 55 years age.

4 Materials and Methods

For the study, a total of 50 respondents belonging to the age group of above 55 years engaged in manual tea plucking activity were selected purposively from three randomly selected tea gardens of Jorhat district of Assam. The data were gathered from the respondents personally by the investigator.

Data on Work Ability Index (WAI) were collected by using the WAI questionnaire. The WAI scores were calculated according to the standard method provided by the Finnish Institute of Occupational Health (FIOH). According to the tool, ‘the main part of the WAI consisted of 7 items, Including current work ability compared with the lifetime best (0–10), work ability in relation to the demands of the job (2–10), number of current diseases diagnosed by a physician (1–7), estimated work impairment due to diseases (1–6), sick leave during the past year (1–5), personal prognosis of work ability 2 years from now (1, 4 or 7) and mental resources, referring to the workers life in general, both at work and during leisure time (1–4). (The number in parentheses for each item indicates the scoring range).’ The WAI Score ranges from 7 to 49

Table 1 Work ability Index of workers engaged in tea plucking activity

WAI Categories	Number	Percentage (%)
Excellent	0	0
Good	0	0
Moderate	16	32
Poor	34	68

points. Based on the WAI, the respondents were grouped according to the following categories: (according to FIOH tool).

Categories	Scores
Poor	:7–27
Moderate	:28–36
Good	:37–43

Frequency, percentage, mean and standard deviations were computed to elicit information according to the objectives of the study.

5 Results and Discussion

5.1 Work Ability Index of Workers Engaged in Manual Tea Plucking Activity

The work ability index (WAI), well-accepted questionnaire-based method to conceptualize work ability, developed by Finnish researchers (Tuomi 1998; Ilmarinen 1985) was used in the present study, which is presented in Table 1.

Mean Work Ability Index of workers engaged in manual tea plucking activity was found 26.96. Higher percentage of workers had poor work ability that is 68%, which is followed by 32% as moderate work ability. No good and excellent work ability was found among the selected workers. This may be the fact that the workers are required to work in an adverse working condition involved more physical activities, which lead to reduced work ability (Fig. 1).

5.2 Dimensions of Work Ability Index (WAI)

For regaining; maintaining and promoting work ability among the workers knowledge on the different dimensions of WAI are necessary. The aim of the present study was the assessment of work ability of the workers engaged in tea plucking activity in tea fields based on the scores obtained from Work Ability Index (WAI). Mean

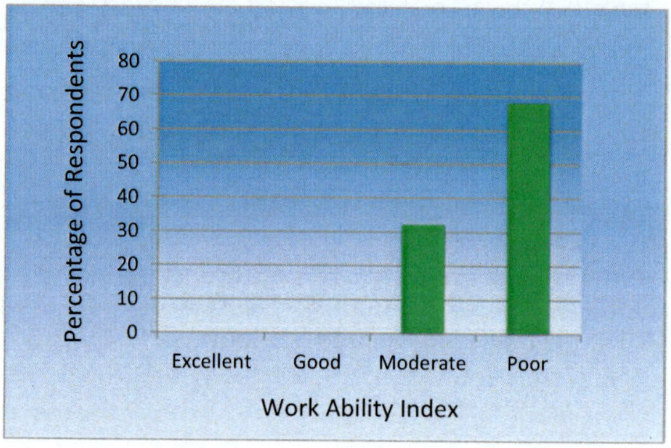

Fig. 1 Work Ability Index of workers engaged in manual tea leaf plucking activity

Table 2 Mean score of WAI dimensions on descriptive variables

Variables	Mean	S.D	Range
1. Current work ability compared with life time best	5.06	1.03	3–10
2. Work ability in relation to demand	5.14	0.7	2–10
3. Number of current disease	1.92	0.87	1–7
4. Estimated work impairment due to disease	3.92	0.56	1–6
5. Sick leave	3.02	0.68	1–5
6. Personal prognosis of work ability 2 year from now	4.9	1.38	1–7
7. Mental resource	3	0.57	1–4
Total WAI score	26.96	5.82	10–49

score of WAI dimensions on descriptive variables was calculated and presented in Table 2.

The data on the use of the first dimension of WAI, termed as 'Work Ability score compared with life time best' (WAS) consist of the worker's self-assessment of his/her current ability compared with the lifetime best (Parameters are according to the tool developed by FIOH). The score ranges from 0 to 10. In the present study, the current work ability was compared with life time best and was found 5.06 among the workers due to the predominance of physical activity had a favourable impact on work ability. Same trend was also observed as regards to work ability in relation to demand of work; which is 5.14. Number of current diseases observed among the workers was 1.92. The diseases more common among the workers were repeated pain in joints or muscle, migraine, infection in urinary tract, skin disease. The estimated work impairment due to disease was found 3.92. As regards to personal prognosis of work ability 2 years from now, it was found 4.9. This may be because that tea garden workers were not much confident. The dimension of mental resources signifies the psychological resources, i.e., enjoying while performing daily tasks and life spirit, positive about the future. Mental resources among the tea workers were found 3.00.

Table 3 Stressors for reducing work ability

Stressors	Mean	S.D	Score range
Musculoskeletal pain	23.4	2.85	6–30
Psychosomatic symptoms	11	1.82	6–18
Ergonomics	7.16	1.00	4–20
Management	7	1.52	3–15
Mental work demand	7.66	0.81	3–15

5.3 Stressors for Reducing Work Ability

Stress at work is a specific type of condition whose sources are at the workplace or in the working environment, is a specific type of condition wherein the work-related factors interact with the human factors in such a way that the person deviates from his/her normal functioning, affecting the work ability. The concept of stress at work comprises changes consequential to stressors present at the workplace over longer periods of time. Stressful jobs may lead to the so-called burnout syndrome, which not only increases the risk of getting ill but also affects private and social life, reduces the sense of self-esteem and the quality of work, as well as safe performance at work (NIOSH 1999).The mean score of the stressors for reducing work ability was calculated and presented in Table 3.

It was revealed that the mean score of musculoskeletal pain is 23.4 out of the score range (6–30).Workers in tea garden have frequent body pain because they have to work in adverse working conditions inherent in the work process, viz., adopting awkward postures reaching for the new shoots to be plucked, which involve long hours of standing, reaching or bending, repetitive movements and carrying heavy loads on back during plucking operation, etc. Same trend was observed as regard to psychosomatic symptoms, which is 11 out of the score range (6–18). Psychosomatic symptoms were studied in terms of tiredness, sleep disturbances, quick or uneven heartbeats, stomach ache, anxiety, etc.

The mean score of ergonomics is found very less that is 7 out of the score range (4–20). Because the tea workers use very awkward postures, and they have to shift from one place to another for plucking leaves by carrying their heavy load on their back. The mean score of management is 8 out of the score range (3–15). Here, the management refers about supervisors help and support, superior's attitude towards worker and considerations of worker's view, which is very bad in tea industry. Superior never considers worker's viewpoint. The mental demand signifies work variation, use of knowledge and skills and need for considerations. Mental demand among the tea workers was found 7.66.

6 Conclusion

The findings of the present study indicated that mean work ability index of workers engaged in manual tea plucking activity was found 26.96. Higher percentage of workers had poor work ability that is 68%, which is followed by 32% as moderate work ability. No good and excellent work ability was found among tea garden workers because they are working in an adverse working condition involved more physical activities, which lead to reduced work ability. The factors responsible for reducing work ability are work-related stressors, long hours of work, with high pace of activity and work demand, physical load, etc. Therefore, it is an utmost necessity to study the working conditions in terms of prevalent risk factors at workplace, so as to enhance the productive performance of the workers along with improving their work ability.

References

Baroowah GP (2006) The legend, life and livelihood of India. LBS Publications, Guwahati, India, pp 75–77

Bhattacharyya N, Baruah M, Farhanaali N (2013) Ergonomic assessment of workers engaged in tea cultivation. Asian J Home Sci 8(2):580–585

Bordia A, Bhardwaj G (ed) (2003) Rethinking pension provision for India. Tata Mcgraw Hill Publishing Company Limited New Delhi

Ilmarinen J (ed) (1985) The ageing. Scand J Work Environ Health 17(1), Supplement 1

Ilmarinen J, Tuomi K, Klockars M (1998) Changes in the work ability of active employees over an 11-year period. Scand J Work Environ Health 23 (Suppl 1):49–57

Ilmarinen JE (2001) Aging workers. Occup Environ Med 58:546–552

Julin H, Lamberg ME, Vinni K, Savolainen H, Niemi ML, Linkola P, Laine A, Kaukinen K, Perimäki R (2001) Tutkimustieto hallinnon välineenä. Työkyvyn ylläpidon tutkimus ja arvioi nti. Raportti 6. Sosiaali- ja terveysministeriö, Kansaneläkelaitos, yöterveyslaitos. Helsinki

Kinsella K, Gist Y (1998) Gender and aging: mortality and health. Bureau of the Census, Economics and Statistics Administration, Suitland, MD. http://www.census.gov/ipc/prod/ib98-2.pdf

Kinsella K, Velkoff VA (2001) An aging world: 2001. National Institute on Aging, Maryland. https://doi.org/10.2307/1534740

Lama MP (1983) Tea industry of darjeeling hill: some aspects of production problem. A paper submitted to a seminar jointly organized by INHERE and ICSSR. New Delhi

Liebig PS, Rajan SI (ed) (2003) An ageing India: perspectives, prospects and policies. The Haworth Press, New York

Sen BB (ed) (2008) Origin and develoment of tea. Guwahati, India, EBH Publishers, pp iii–v

Shettar SC (2013) Problems of aged in changing Indian scenario. Yojana Archives

Tuomi K, Ilmarinen J, Jahkola A, Katajarinne L, Tulkki A (1998) Work ability index, 2nd edn. Finnish Institute of Occupational Health, Helsinki, Finland

United Nations (2002) United Nations principles for older persons. United Nations, United Nations Human Rights, Office of the High Commissioner

United Nations (2011) World population ageing, 1950–2050. Department of Economic and Social Affairs, Population Division, New York

World Bank (1984) From poor areas to poor people: China's evolving poverty reduction agenda. World Bank, Washington, DC

World Bank (2001) India: the challenge of old age income Security, finance and private sector development, South Asia Region, Report 22034-In, Washington. World Health Organisation (1985)

Chapter 2
Ergonomic Assessment of Lower Back Pain Among Coir Industry Workers and Workstation Modification

M. Satheeshkumar and K. Krishnakumar

1 Introduction

Work-related musculoskeletal disorders (WMSDs) are injuries that are caused by workplace activities, which affect the musculoskeletal system of the human body (Punnett 2004). The musculoskeletal system consists of muscles, tendons, ligaments, joints, peripheral nerves, and supporting blood vessels. The most common body regions affecting the WMSDs are lower back, neck, shoulder, and hand (Putz-Anderson et al. 1997). Awkward trunk posture due to poor workstation layout be the reason for pain and disorders in different body regions among workers (Keyserling et al. 1988). The studies on WMSDs are reported by a number of researchers from different countries in many fields (David 2005, Da Costa and Vieira 2010, Osborne et al. 2012). In several articles, it is mentioned that lower back is the most prevalent WMSDs affected body region (Marras 2000). In a review paper by Da Costa, it is established that the main risk factors of lower back pain (LBP) were awkward working postures (Da Costa and Vieira 2010). NIOSH a health organization in the United States reported that an awkward working posture is an important factor of risk in the development of LBP. A number of reviews of the published literatures have reported that trunk flexion and rotation are major reasons for LBP (Hoogendoorn et al. 2000). In a study of prevalence of WMSD in coir products industry workers, it is conveyed that the lower back is the most affected body region (58.8%) (Satheeshkumar and Krishnakumar 2018). Several research articles were published since the last 30 years concerned with problem of LBP due to physical workload. Laura Punnett et al. studied about the LBP and awkward trunk postures in the workers of automobile assembly workstation and concluded that occupational posture requirements in workstation are important causes of chronic back disorders.

M. Satheeshkumar (✉) · K. Krishnakumar
Department of Mechanical Engineering, CET, Thiruvananthapuram, India

K. Krishnakumar
e-mail: kkk@cet.ac.in

© The Author(s), under exclusive license to Springer Nature Singapore Pte Ltd. 2022, corrected publication 2022

D. Bijulal et al. (eds.), *Technology-Enabled Work-System Design*, Design Science and Innovation, https://doi.org/10.1007/978-981-16-1884-0_2

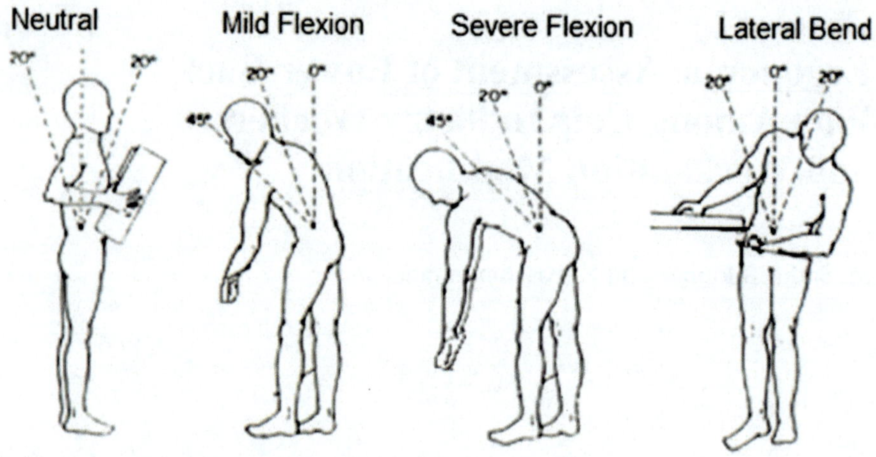

Fig. 1 Standard trunk posture classification. *Source* (Punnett et al. 1991)

Anthropometric measurement and working posture of workers are to be considered in the design of tasks, tools, and workplaces for eliminating the WMSDs (Punnett et al. 1991). Hoogendoorn Wilhelmina et al. conducted a study among 861 workers with no LBP of 34 industries in Netherlands to examine the association between flexion, rotation of the trunk, lifting work and the occurrence of LBP. The paper reported that the body postures such as twisting, trunk flexion and lifting are the most frequently reported physical risk factors for the cause of LBP (Hoogendoorn et al. 2000). The trunk posture that causes LBP was classified in a paper by Laura Punnett et al., the classifications were defined as 'neutral' when the trunk bends forward or twist in any direction at an angle less than 20°, 'mild' when flexion the trunk bend forward more than 20° and less than 45°, 'severe' when the trunk bends forward more than 45° and lateral bending or twisting more than 20°, which is depicted in Fig. 1 (Punnett et al. 1991).

In a preliminary investigation, the pneumatically operated coir mat weaving machine under study, the trunk flexion angle measures more than 26° and lateral bend more than 24°. This posture of workers at this workstation may be the reason for back pain in the workers. Moreover, the popular ergonomics assessment tools like Rapid Upper Limb Analysis (RULA) and Ovako Working posture Assessment System (OWAS) score received are very high value. As per the given guidelines of RULA and OWAS tools, change in the posture is required immediately. Usually, the risk of lower back injury has been estimated using the analysis of peak loading conditions observed during a task.

More recently, the risk of LBP from cumulative loading experienced over the course of the work shift has been documented. Cumulative lower backloading is the accumulation of all loads (both high and low) occurring over the course of a work shift. Research has indicated that there is most likely an association between cumulative spinal loading and LBP (Waters et al. 2006).

Digital human modeling (DHM) software is the popular tool for ergonomic evaluation of new product designs and workstation design and modifications. An article by Chaffin describes the development and applications of digital human models in the fields of ergonomics up to the year 2008, which gives an overall idea about the application DHMs for workplace design (Chaffin 2008). An article by Naumann describes the application of DHM in the field of product design, assembly and manufacturing. The paper highlighted the application of DHMs in workplace ergonomic assessment, hand object orientation, study of eye movement, and comfort assessment (Naumann and Roetting 2007). An article by Alvarez-Casado et al. reported the modification of the workplace in a small fishing vessel for preventing WMSDs problems of fisherman in Spain. The working posture of fisherman was assessed and modified with the DHM software, so as to avoid awkward working postures and biomechanical stresses at lower back (Alvarez-Casado et al. 2016). An article by Lars Fritzsche reported an ergonomic study with DHMs for analyzing certain working static postures of workers in a car assembly line. Author also reported that DHM simulation provides good performance in the ergonomic evaluation, but harder to estimate action forces (Fritzsche 2010). An article by Sougata Karmakar, vision analysis was conducted with digital human modeling software and analyzed with vision tools like view cones, blind spot area, eye view windows, reflection zone, etc. In the article, the author pointed out that vision investigation tool of DHM software is very promising tool for the assessment of alignment and position of various displays and control nobs in the cockpit (Karmakar et al. 2012). An article by Chnag et al. proposes a method for conducting workplace assessment for the prevention of WMSDs. In this work, workplace motion data, biomechanical assessment, and posture assessment are conducted in automobile assembly task, and improvements have been implemented (Chang et al. 2007).

In this work, an ergonomic study was conducted on a coir mat weaving machine to evaluate the LBP caused by the awkward trunk postures like bending and twisting of workers. This paper also deals with modification of weaving workstation to get a better working posture so as to reduce the LBP of the worker.

The coir mat weaving machine is made up of wooden or metal frames with provision for setting the warp and weft and is purely manually operating. Warping and wefting constitute the two basic elements in the weaving process. The lengthwise coir warp is aligned systematically on a frame or loom while the crosswise coir weft is inserted between the warp plies with the help of a bobbin carrier called shuttle. The modern coir mat weaving machines constitute a metal frame with all parts as like the traditional machine and it is pneumatically operated. In this paper, the work system selected for the study is a modern weaving machine. These machines were manufactured by small-scale industries, and little attention was given to the ergonomic principles in the design.

2 Materials and Methods

A CAD model of the weaving workstation (loom) is created with the Solid Works software. The actual dimension is collected from the industry, and it is used for the CAD drawing. The existing loom is shown in Fig. 2.

The Tecnomatix Jack (Version 8.0), the DHM-based simulation software for ergonomic assessment is used in this study. The CAD model is inputted to the Jack software environment for the ergonomic assessment. The 3D CAD drawing of the work system and digital human model in Jack environment is shown in Fig. 3.

The popular ergonomics assessment tools, RULA and OWAS, are used in this study. The upper body risk factors are assessed with RULA ergonomic tool. RULA and OWAS analysis were carried out in male and female population of size 5th, 50th, and 95th percentile with the Asian-Indian anthropometric database available in the software.

The actual work system when a worker engaging the weaving work is captured by a video camera. The time required for different work elements in a work cycle was noted by playing the video in slow motion. The different task elements are (a) Get the shuttle, (b) Insert the shuttle on the right side of the warp, (c) Get the shuttle from the left side, (d) Put the shuttle on the platform, (e) Tighten the weft, (f) Operate the pedals, (g) Operate the air pressure knob. The average number of tasks per 8 h shift was received from the video. Simulation was carried out with the task simulation builder (TSB) module available in Tecnomatix Jack software, which provides estimates of peak lower back compressive force at the L4/L5 joint. In TSB simulation module, the movements of the worker and objects are created with the exact time taken for each task element. From the peak lower back compressive force, the cumulative load per shift is calculated with the formula given below.

Fig. 2 Actual work system in a coir industry

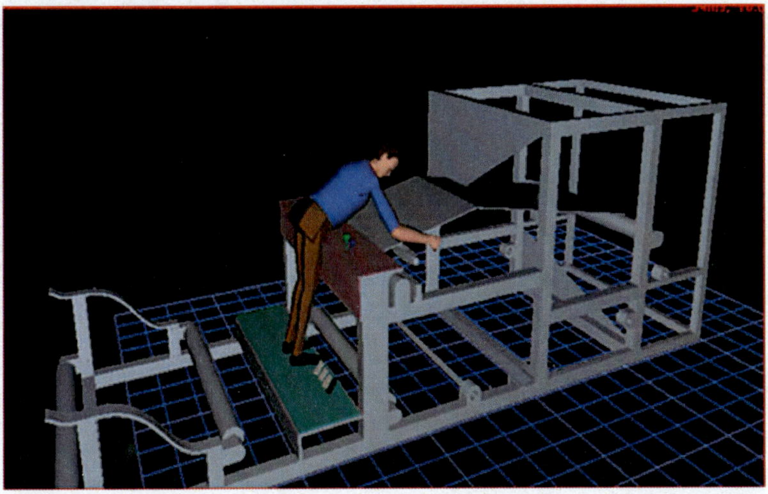

Fig. 3 Digital human model and the work system in Jack Software

$$CL_{Shift} = \left(\sum^{N} F_n \cdot t_n\right) \times k \tag{1}$$

CL_{shift}—Cumulative load per shift (Newton second-Ns).
F—Peak compressive force during the task element (Newton-N).
t—Action time for the task element (second-s).
N—Number of elements in a task.
k—Average number of tasks per 8 h work shift ($k = 3500$).

3 Results

The existing weaving workstation (loom) was modeled in the Jack software, and simulation was conducted, then the RULA and OWAS score of this system received are 7 and 3, respectively. As per the RULA guidelines, this value is not acceptable and action required is 'further examination and posture modifications are the immediate requisite'. In the existing loom, the warp is arranged horizontally, that is the warp angle is 0°. Initially, simulation was inducted with digital human of size 95th percentile Jack male one. Next, the CAD model of the work system is altered by changing the warp angle dimension by 10° inclined to horizontal and then the simulation is repeated and gets RULA and OWAS score. In this way, the simulation is continued up to 80° warp angle with 10° increment in each step. The RULA and OWAS scores for 0–80° are shown in Fig. 4.

The experiment was repeated with digital human of size 50th male, 05th male, 95th female, 50th female, and 05th female percentiles. From the experiments, we got

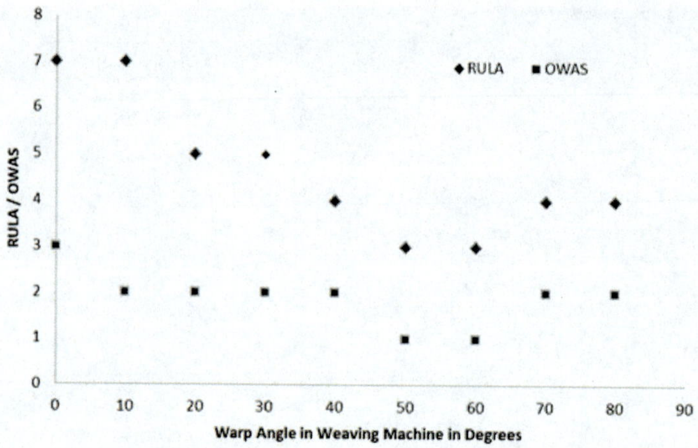

Fig. 4 RULA and OWAS scores of 95th male

minimum RULA and OWAS scores corresponding to the warp angle between 50 and 60° for all class of populations. From the results, we concluded that the optimum value of warp angle is 60°. The optimized structure with warp angle 60° and the corresponding human posture are depicted in Fig. 5.

The simulation experiment is conducted in the existing work system with digital humans of the size 95th male, 50th male, 05th male, 95th female, 50th female, and 05th female. The angles like trunk flexion, axial rotation, and lateral rotation are

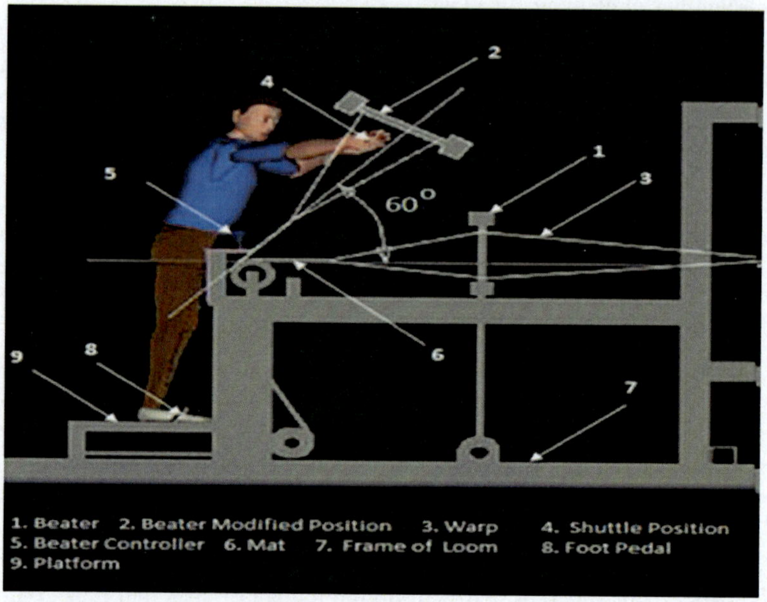

Fig. 5 Schematic Diagram of the Modified Work system

Table 1 Trunk posture angles in existing and modified workstation

Trunk posture angle	Existing (warp angle 0°)		Modified (warp angle 60°)	
	Minimum	Maximum	Minimum	Maximum
Trunk flexion	26°	46°	9°	33°
Axial rotation	15°	22°	6°	20°
Lateral rotation	24°	34°	5°	32°

Table 2 Cumulative load in actual per shift and modified work system

Gender	Human size (Percentile)	Cumulative load (MNs)/shift		
		Actual	Modified	Reduction (%)
Male	95th	24.587	18.109	26.35
	50th	25.634	23.009	10.24
	05th	13.877	12.785	07.87
Female	95th	17.717	15.330	13.47
	50th	15.473	14.616	05.54
	05th	14.630	13.797	05.69

measured for all the populations mentioned above. The minimum and maximum values obtained for trunk flexion angle are 26 and 46° that for axial rotation angle are 15 and 22°, and that for lateral rotation angle 24 and 34°. Based on the trunk posture classification by Laura Punnett et al., the working posture of the human who engaged in the work system under study can be classified into mild or severe class. Hence, we resolved that the requirement of work system modification is necessary to reduce WMSDs. By modifying the existing work system with a warp angle of 60°, the minimum and maximum values of trunk flexion, axial rotation, and lateral rotation angles are optimized as shown in Table 1.

The cumulative load per shift for the existing and modified work system is calculated with Eq. 1, and the values are given in Table 2.

About 5.54–26.35% reduction in cumulative load per shift was received when the work system is modified, and this reduction is a very considerable amount.

4 Conclusion

This paper illustrated the simulation of work system with the use of digital human model. The cumulative lower backloading is assessed, and we suggested modification in design of work system for getting a better working posture. We conclude that it is possible to minimize the awkward working posture and WMSD problems with ergonomic modifications of the work system. The results received and the ergonomic benefits along with our conclusion are communicated with the manufacturer of the machine for the modification of the machine. It is observed that most of the work systems in coir industry are not following the ergonomic principle; so ergonomic intervention is essential for all other work systems also.

Acknowledgements The authors accord the deepest sense of gratitude to the authorities of coir manufacturing company for providing the facility and to the workers for their sincere involvement in data collection throughout this study.

References

Alvarez-Casado E et al. (2016) Using ergonomic digital human modeling in evaluation of workplace design and prevention of work-related musculoskeletal disorders aboard small fishing vessels. Hum Fact Ergon Manuf Serv Ind 26(4):463–472

Chaffin D (2008) Digital human modeling for workspace design. Rev Hum Fact Ergon 4:41–74

Da Costa BR, Vieira ER (2010) Risk factors for work-related musculoskeletal disorders: a systematic review of recent longitudinal studies. Am J Ind Med 285–323

David GC (2005) Ergonomic methods for assessing exposure to risk factors for work-related musculoskeletal disorders. Occup Med 55(3):190–199

Punnett LDH (2004) Wegman: work related musculoskeletal disorders the epidemiologic evidence and debate. J Electromyogr Kinesiol 13–23

Chang SW, Wang MJJ (2007) Digital human modeling and workplace evaluation: using an automobile assembly task as an example. Hum Fact Ergon Manuf Serv Ind 17(5):445–455

Fritzsche L (2010) Ergonomics risk assessment with digital human models in car assembly: simulation versus real life. Hum Fact Ergon Manuf Serv Ind 20(4):287–299

Hoogendoorn WE, Bongers PM, de Vet HC, Douwes M, Koes BW, Miedema MC, Bouter LM, Hoogendoorn WE, Bongers PM, de Vet HC (2000) Flexion and rotation of the trunk and lifting at work are risk factors for low back pain: results of a prospective cohort study. Spine 25(23):3087–3092

Karmakar S et al. (2012) Application of digital human modeling and simulation for vision analysis of pilots in a jet aircraft: a case study. Work 41(Supplement 1):3412–3418

Keyserling WM, Punnett L, Lawrence J (1988) Trunk posture and back pain—identification and control of occupational risk factors. Appl Ind Hyg 87–92

Punnett L, Fine LJ, Keyserling WM, Herrin GD, Chaffin DB (1991) Back disorders and non-neutral trunk postures of automobile assembly workers. Scand J Work Environ Health 337–34

Marras WS (2000) Occupational lower back disorder causation and control. Ergonomics 43(7):880–902

Naumann A, Roetting M (2007) Digital human modeling for design and evaluation of human-machine systems. MMI Interaktiv

Osborne A, Blake C, Fullen BM, Meredith D, Phelan J, McNamara J, Cunningham C (2012) Prevalence of musculoskeletal disorders among farmers: a systematic review. Am J Ind Med 55(2):143–158

Putz-Anderson V, Bernard BP, Burt SE, Cole LL, Fairfield-Estill C, Fine LJ, Nelson N (1997) Musculoskeletal disorders and workplace factors. National institute for occupational safety and health (NIOSH), vol 104

Satheeshkumar M, Krishnakumar K (2018) Chapter 10; study on work-related Musculoskeletal disorders among coir industry workers in the state of Kerala, India. Ergonomic design of products and work system—21st century. Springer, pp 117–130

Waters T, Yeung S, Genaidy A, Callaghan J, Barriera-Viruet H, Deddens J (2006) Cumulative spinal loading exposure methods for manual material handling tasks; Part 1: is cumulative spinal loading associated with lower back disorders? Theor Issue Ergon Sci 113–130

Chapter 3
Postural Assessment While Performing a Drilling Task Using the Kinect Sensor for a Hand Drilling Machine

M. Shah Faizan, Z. Shah Fahad, and Mohammad Muzammil

1 Introduction

The advent of globalization led companies to manufacture products in an excessive quantity with better quality, in a minimum amount of time possible. Despite safety guidelines in the workplace, the workers need to fulfill this challenge in an unergonomic work environment with long working hours, especially in developing countries. This led the workers on the verge of health issues and worsen their quality of life. Some other task-based factors such as static and awkward posture, forceful exertion, vibration, and repetition further laid challenges and cause body and joint discomfort. The presence of discomfort for a long period will lead to the development of musculoskeletal disorder (MSD) and injuries.

In MSD, the soft tissues like muscles, joints, and tendons get ruptured. It ceases to work in its natural way and offers difficulties in performing the basic activities (Safety et al. 2007). The MSD in workers is one of the main concerns for industrialists that causes absenteeism and lower down the productivity of industries. One of the surveys revealed that around 10% of the US workforce reported MSD (Bureau of Labor Statistics 2007). The primary indication for the development of MSD is the lower back pain that often occurs due to an awkward posture with intense or repetitive lifting (Garg 1989). The incidence of awkward posture and MSD is more severe in medium and small-scale industries because of the lack of ergonomics guidelines and their implementation in the workplace. The research showed that the unergonomic working condition often resulted in lower productivity, poor product quality, illness, and sometimes injuries (Damaj et al. 2016). Meanwhile, designing the workplace with ergonomic guidelines prevents the occurrence of work-related accidents by 50% (Gibb et al. 2004).

M. S. Faizan (✉) · Z. S. Fahad · M. Muzammil
Mechanical Engineering Department, Aligarh Muslim University, Aligarh, India

© The Author(s), under exclusive license to Springer Nature Singapore Pte Ltd. 2022, corrected publication 2022

D. Bijulal et al. (eds.), *Technology-Enabled Work-System Design*, Design Science and Innovation, https://doi.org/10.1007/978-981-16-1884-0_3

Maintaining the proper working posture in the workplace is the best initiation of applying the ergonomic guidelines. The proper working posture enhances productivity and minimizes the accidents and risk of MSD (Widanarko et al. 2012). Proper working posture can be described as the orientation of body segments in relation to each other that results in the minimum discomfort of the limbs while performing the task (Haslegrave 1994). Working posture is one of the best estimators to measure the inaptness involved that is described as the deflection from the static posture (Westgaard and Aaras 1984). Moreover, it is also used to estimate the amount of comfort/discomfort associated and the effort required with the particular task (Khan and Muzammil 2018). In the manufacturing industries, grinding, milling, turning, and drilling are the tasks that are mostly afflicted with the poor working posture, among which drilling has the most critical working posture that needs attention.

Drilling is one of the most important and widely used processes in multiple sectors such as the oil and gas sector, construction sector, production sector, and the automotive sector. It is usually linked with vibration, repetition, forceful action, awkward, and standing posture. According to an estimate of around 17.6%, injury and accident cases were associated with the use of a powered drill (Aghazadeh and Mital 1987). According to the research, tasks that are associated with forceful exertion have high chances of the development of MSD (Veddar and Carey 2005). Moreover, tasks accompanied by awkward posture, forceful exertion, and repetition led to the high chances of repetitive strain injury and MSD (Escorpizo 2008). The posture maintained during the performing of a drilling task significantly affects the worker's well-being, therefore, efforts are required to minimize MSD by analyzing and providing the proper working posture.

In this paper, the posture of a worker under different hand drilling positions is analyzed. The OWAS application incorporated with the Kinect sensor is employed to assess the posture. The study aims to estimate the discomfort associated with different postures and to know its variation with the change in posture.

2 Method

2.1 Kinect Sensor

Kinect is launched initially as the gaming equipment by Microsoft but later used in research to record the movement of a person in 3D space. The sensor is equipped with the RGB camera, and the infrared camera and projector, a combination of which help the sensor to measure the person's position precisely. The Kinect tracks human motion by detecting the major limb of the body using an algorithm and gave its output in the form of 3D joint coordinates (Figlah et al. 2015). The low cost, usability, extended availability, and the portability of the Kinect make it an appropriate option to be used in the research. Moreover, its capability to record without attaching any pointers to the body makes it even more attractive to carry out the postural assessment in industrial applications (Manghisi et al. 2017; Mas and Marzal 2014).

2.2 *Ergonautas Software*

Ergonautas is an intermediate software between the Kinect and the OWAS application. It retrieves the information from the Kinect, which is in the skeletal data form, and forward it to the OWAS application in the form of joint coordinates. The Ergonautas automatically modified the data obtained from the Kinect, to coincide with the coordinates of feet to that of ground. The software is free to download with a 16-day trial from the mentioned link (http://www.ergonautas.com/lab/kinect).

2.3 *OWAS Application*

The OWAS application is a computerized tool within the Ergonautas software, run on the Ovako Working Posture Analysis System-based algorithm. It calculates various angles between the body segments using the data of 3D coordinates of 20 joints, obtained previously from the Kinect. The information about various angles is then used to estimate the orientation of arms, trunk, and legs. The body posture is marked with four digits code, each digit indicates the orientation of arms, trunk, legs, and the load carried by the worker. Depending upon the code allotted, each posture is categorized into a single class out of four classes of risk as shown in Table 1. Further, after summarizing the risk status of all the postures within a task, a global risk rating is assigned, which indicates the intensity of risk involved in the task.

Table 1 Risk categories of OWAS Method

Category 1	Postures have no harmful effect
	No workplace modification or change in posture is needed
Category 2	Postures have some harmful effects
	Workplace modification or change in posture is needed in future
Category 3	Postures have clear harmful effects
	Workplace modification or change in posture is needed as soon as possible
Category 4	Postures have considerable harmful effects
	Workplace modification or change in posture is needed immediately

2.4 Subjects

Ten healthy subjects were recruited to perform the experiment. All subjects were male, and simple random sampling is used for the selection of the subjects. They were briefed about the experiment and mildly trained on a hand drilling machine. All the subjects are healthy and did not hold any back or leg injury during the performing of the task.

2.5 Experimental Setup

A hand drilling machine (Kobalt Electric Drill KDM-13,) was used to carry out experimental investigations. A wooden board and HSS drill bit with a diameter of 6 mm were selected for a drilling operation. Fifty postures were captured using Kinect for each condition, with a rate of 5 frames/sec. The field range of the Kinect was set between 2 and 3 m. Since the task involved load lifting in the form of hand drilling machine (weight: 1.8 kg), therefore the load handled was input to be below 10 kg in an OWAS application.

2.6 Task Description

Subjects were instructed to carry out drilling using hand drilling machine while maintaining the five postures based on drilling location, namely: above the shoulder height, at the shoulder height, below the shoulder height, overhead, and stooped posture. For each posture maintained, the drilling is performed on three reach conditions, namely close, middle and far, thus, evaluating the posture on 150 conditions (10 subjects × 5 postures × 3 reaches).

2.7 Data Analysis

The Kinect recorded the worker's posture as shown in Fig. 1, and transferred the data to the Ergonautas. Ergonautas together with the OWAS application gave the level of risk involved in arms, trunk, and legs. The data of the risk involved from all the 10 subjects are then collected, averaged and analyzed, shown in Table 2.

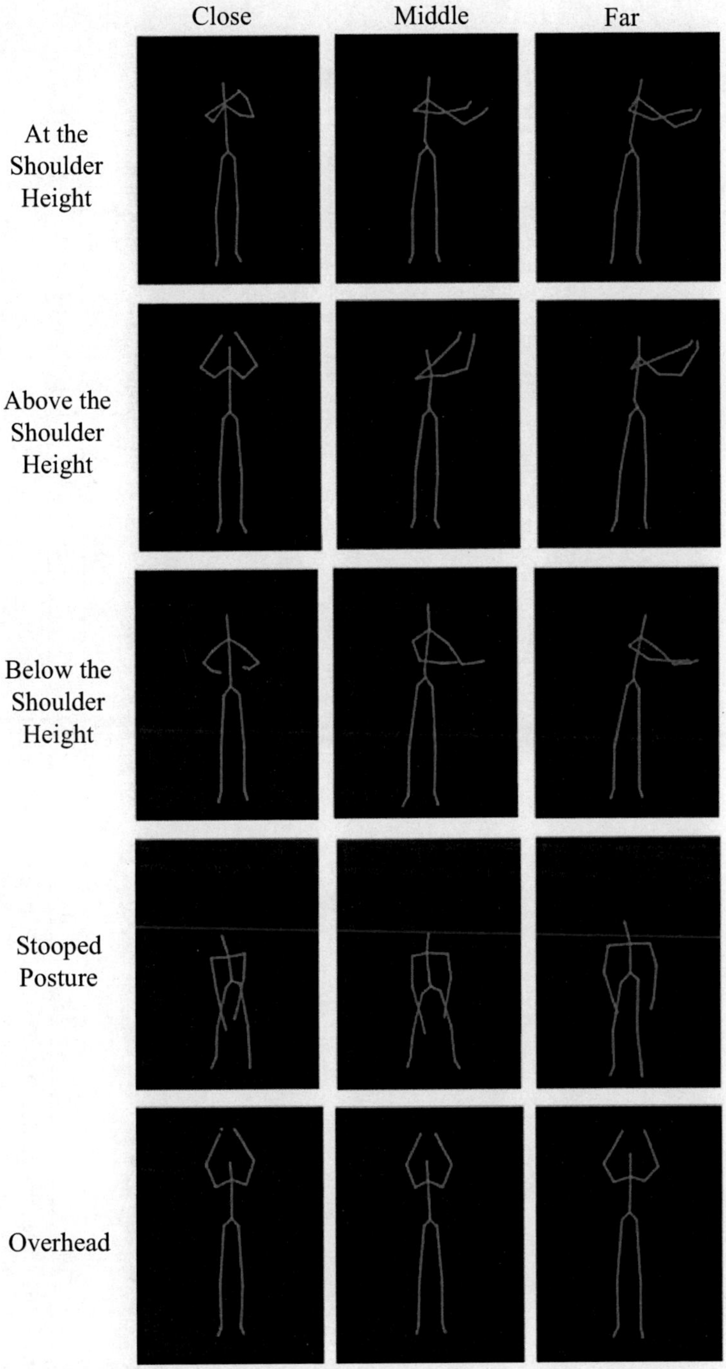

Fig. 1 The posture of a worker for different drilling positions and reach conditions

Table 2 The member's global risk and overall risk degree associated with the worker's posture

Experiment conditions	Posture	Reach	Posture frequency				Member's global risk			Global risk rating	Risk degree
			Risk 1	Risk 2	Risk 3	Risk 4	Trunk	Arms	Legs		
1	At the shoulder height	Close	12	30	8	0	SH	SH	SH	2.12	SH
2		Middle	8	33	9	0	SH	SH	SH	2.26	SH
3		Far	4	35	11	0	SH	SH	SH	2.35	SH
4	Above the shoulder height	Close	7	22	18	3	SH	DH	SH	2.72	DH
5		Middle	4	21	21	4	SH	DH	SH	2.81	DH
6		Far	3	18	23	6	SH	DH	SH	2.89	DH
7	Below the shoulder height	Close	21	24	5	0	SH	AA	SH	1.76	SH
8		Middle	18	26	6	0	SH	AA	SH	1.82	SH
9		Far	19	23	8	0	SH	SH	SH	1.95	SH
10	Stooped posture	Close	1	12	27	10	DH	AA	SH	3.42	DH
11		Middle	0	8	29	13	EH	AA	SH	3.54	EH
12		Far	0	2	33	15	EH	SH	SH	3.63	EH
13	Overhead	Close	0	7	31	12	SH	EH	SH	3.57	EH
14		Middle	0	10	26	14	SH	EH	SH	3.61	EH
15		Far	0	5	28	17	SH	EH	SH	3.67	EH

AA - Acceptable, SH - Slightly Harmful, DH - Distinctly Harmful, EH - Extremely Harmful

3 Results

The member's global risk and overall global risk associated with worker's posture during hand drilling tasks are summarized in Table 2. The results showed that the worker's posture during overhead drilling with far reach condition was under maximum risk with a global risk rating of 3.67. However, the worker's posture while drilling below the shoulder height at close reach condition involves a minimum risk with a global risk rating of 1.76. It is also evident from the results that the stooped posture, drilling above the shoulder height and at the shoulder height was under the risk category of extremely harmful, distinctly harmful and slightly harmful, respectively.

The member's global risk for the trunk was extremely harmful during drilling at a stooped posture. On the other hand, the member's global risk for arms was distinctly harmful and extremely harmful for drilling above the shoulder height and overhead drilling conditions, respectively. The results showed that out of 50 postures captured, overhead drilling has a maximum number of risky postures (i.e. collectively in risk categories 3 and 4) while drilling below the shoulder height has a minimum. It was also observed from results that the global risk rating, member's global risk, and risky postures increase as the reach condition shifts from close to far.

4 Discussion

The study showed that the drilling at the shoulder height and below the shoulder height has postures in the risk categories 1 and 2. This is because the angle at the shoulder joint is smaller, leads to lesser muscle strain and risk. Hence, for these working positions, the posture has no or very slight harmful effect on the musculoskeletal system, so workplace modification or change in posture should be made in the upcoming plans. Moreover, drilling above the shoulder height has the majority of posture under risk categories 2 and 3, this is because the hand is in the overhang condition for the majority of the time, that further causes muscle strain. Therefore, workplace modification or change in posture must be made as soon as possible. The stoop and overhead drilling have the majority of posture under the risk categories 3 and 4. The reason behind it is the continuous bending of the spine in the case of stoop posture, which leads to the fatigue of spine tissue and causes discomfort at the back. For overhead drilling, the high elevation of the arm causes a large angle at the shoulder joint that leads to muscle strain and fatigue of the hand. Hence, the working posture is ergonomically not appropriate and workplace modification or change in posture must be made immediately. Thus, the working posture during hand drilling demands modification to make it ergonomically more acceptable. The modification in the working posture can be either provided through the incorporation of ergonomic interventions or the use of assistive devices. The study also showed that the far reach position was associated with more risky posture as compared to the middle or close reach position, whereas the close reach position has the minimum number of risky postures.

However, the posture captured using the Kinect hold certain limitations. The accuracy of the Kinect sensor in detecting the posture decreases as the sensor moves away from the frontal plane, longitudinally as well as transversely. Moreover, the use of the Kinect sensor is not recommended for long-duration study as it gets heated up after a certain time.

5 Conclusion

In this study, the worker's posture was assessed for different drilling positions and under different reach conditions. It is evident from the study that the worker's posture during stoop above the shoulder height and overhead drilling is harmful and may cause a detrimental effect on the worker's musculoskeletal system. The study also revealed that it is better to perform drilling with a close reach to avoid poor posture, lessen the muscle strain and minimize the risk of MSDs during the drilling.

Acknowledgements The author would like to thank "Ergonautas Lab, Universidad Politecnica de Valencia" for providing the Ergonautas Software and giving a license to use it in the assessment of a worker's posture.

References

Aghazadeh F, Mital A (1987) Injuries due to hand tools. J Appl Ergon 18(4):273–278

Bureau of Labor Statistics (2007) Employed persons by detailed industry, sex, race, and hispanic or latino ethnicity. http://www.bls.gov/news.release/empsit.nr0.htm. Accessed 09 Nov 2018

Damaj O, Fakhreddine M, Lahoud M, Hamzeh F (2016) Implementing ergonomics in construction to improve work performance. In: 24th Annual conference of the international group for lean construction. Boston, pp. 53–62

Escorpizo R (2008) Understanding work productivity and its application to work-related musculoskeletal disorders. Int J Ind Ergon 38(3–4):291–297

Figlah N, Cihan A, Esen H, Figlah A, Cesmeci D, Gullu MK, Yilmaz MK (2015) Image processing-aided working posture analysis: I-OWAS. J Comput Ind Engi 85:384–394

Garg A (1989) Epidemiological basis for manual lifting guidelines. NIOSH—937384

Gibb A, Haslam R, Hide S, Gyi D (2004) The role of design in accident causality. In: Proceedings from a research and practice symposium. University of Oregon Labor Education and Research Center, Eugene

Haslegrave CM (1994) What do we mean by a working posture. J Ergon 37(4):781–799

Khan MA, Muzammil M (2018) Design and evaluation of a modified drilling method. Int J Ind Ergon 67:114–122

Manghisi VM, Uva AE, Fiorentino V, Trotta GF, Monno G (2017) Real-time RULA assessment using kinect v2 sensor. J Appl Ergon 65:481–491

Mas JAD, Marzal JA (2014) Using kinect sensor in observational methods for assessing postures at work. J Appl Ergon 45:976–985

The Occupational Safety and Health Act (2007). http://www.health.go.ke/wp-content/uploads/2015/09/OSH%20Act%202007.pdf. Accessed 10 Nov 2018

Veddar J, Carey E (2005) A multi-level systems approach for the development of tools, equipment and work processes for the construction industry. J Appl Ergon 36:471–480

Westgaard RH, Aaras A (1984) Postural muscle strain as a causal factor in the development of musculoskeletal illness. J Appl Ergon 15(3):162–174

Widanarko B, Stephen L, Stevenson M, Devereux J, Cheng S, Pearce N (2012) Prevalence and work-related risk factors for reduced activities and absenteeism due to low back symptoms. J Appl Ergon 43:727–737

Chapter 4
Determination of Lifting Index for Paddy Storage Activity Performed by Farm Women

M. Kalita, R. Borah, and N. Bhattacharyya

1 Introduction

Handling of material manually is the most predominant and economical mode of activity in our daily life. There are a number of variations in the manner in which a load is being handled. It may be lifted from the floor or from higher or lower level than the floor. Moreover, the load lifted may be fixed or variable at times (Gangopadhyay et al. 2003**)**. Forces produced during various processes of material handling, e.g. pulling, pushing, lifting, carrying etc. may sometimes be much above the normal tolerance level of the worker. Storage of paddy grains is one of the most drudgery-prone and manual handling postharvest activities, which is predominantly performed by farm women in Assam. Storage of paddy grains is performed by more than 70% of the rural Assamese women. After proper sun drying, the paddy is stored in storage structures such as *bhoral*, *mer*, *duly*, gunny bags, etc. for consumption in the next year or for commercial purposes. *Bamboo basket* is a conventional tool used for carrying the paddy from yard to the place of storage structures. The farm woman usually carried more than 16 kg of load of paddy on their waist side at a time by adopting an awkward posture The *bamboo basket* used by farm women for paddy storage activity was not in conformity to the body types and anthropometric measurements. There is no provision of base and grip for handling the tool, which increases the grip fatigue of the farm women during storage.

Musculoskeletal disorders (MSDs) are important causes of work incapacity and loss of workable life of farm women in storing activity. For minimizing the musculoskeletal disorders (MSDs) of farm women in paddy storage activity, determination of permissible weight limit and lifting index (LI) is very essential. Keeping this in mind, the study was carried out to determine the recommended weight limit (RWL)

M. Kalita (✉) · R. Borah · N. Bhattacharyya
Dept. of Family Resource Management and Consumer Science, College of Community Science, Assam Agricultural University, AAU, Jorhat-13, India

© The Author(s), under exclusive license to Springer Nature Singapore Pte Ltd. 2022, corrected publication 2022

D. Bijulal et al. (eds.), *Technology-Enabled Work-System Design*, Design Science and Innovation, https://doi.org/10.1007/978-981-16-1884-0_4

and lifting index (LI) of farm women in paddy storage activity by using conventional basket.

2 Methodology

The present study was conducted in Jorhat subdivision of Jorhat district, Assam. Two development blocks, namely, North West Development Block, Dhekorgorah, and Jorhat Development Block, Bagh-choong were purposively selected for the study. Thirty farm women in the age group of 25–35 years who had been involved in paddy storage activity for last 10 years were selected for the purpose of the study. The farm women who were non-pregnant, non-lactating, and having normal blood pressure without any major illness were selected for the sample. Lifting index (LI) is used for determining recommended weight limit (RWL) and lifting index (LI) of farm women for paddy storage activity. For assessing recommended weight limit (RWL) of farm women and lifting index (LI) in paddy storage activity, the revised NIOSH lifting equation was applied (Nelson et al. 1994/2010).

Lifting index (LI) is the ratio between the load actually lifted and the recommended weight limit (RWL). Lifting index less than 1 indicates workers are not in the risk of injury and an index more than 1 indicates workers are at risk, and design modification of the conventional basket is necessary. The goal of NIOSH in developing the equation is to protect workers by setting limits on manual lifting.

Recommended weight limit (RWL) = (LC) (HM) (VM) (DM) (AM) (FM) (CM).

Equation in Metric.

- LC = load constant = 23 kg
- HM = horizontal multiplier = 25/H
- VM = vertical multiplier = [1—(0.003 |V-75|)]
- DM = distance multiplier = [0.82 + (4.5/D)]
- AM = asymmetric multiplier = [1—(0.0032A)]
- FM = frequency multiplier
- CM = coupling multiplier

There are four different steps for using NIOSH lifting equation for manual material handling tasks. These are as follows:

Step 1: Measure and record lifting task variables.

Lifting tasks	NIOSH lifting variables								
	H Horizontal location (10–25)	V Vertical location (0–70″)	D Travel distance (10–70)	A Angle of asymmetry (0–135°)	C Coupling (1 = good, 2 = fair, 3 = poor)	F Frequency (0.2–15 lifts/min)	L Avg. load lifted (lbs.)	Maximum load lifted (lbs.)	Duration (1, 2, 8 h)

Step 2: To determine the multipliers to use in the NIOSH equation.

	Multipliers (in metric)					
Lifting tasks	HM = horizontal multiplier = (25/H)	VM = vertical multiplier = [1—(0.003 |V—75|)]	DM = distance multiplier = [0.82 + (4.5/D)]	AM = asymmetric multiplier = [1—(0.0032A)]	FM = frequency multiplier (from Table 5 in Appendix A)	CM = coupling multiplier (from Table 7 in Appendix A)

Step 3: To calculate the NIOSH recommended weight limit (RWL).

$$\text{RWL} = (51)\,(\text{HM})\,(\text{VM})\,(\text{DM})\,(\text{AM})\,(\text{FM})\,(\text{CM})$$
$$\text{Or}(23)\,(\text{HM})\,(\text{VM})\,(\text{DM})\,(\text{AM})\,(\text{FM})\,(\text{CM})$$

Step 4: The final step is to calculate the lifting index (LI)

$$\text{LI} = \frac{\text{Object weight}}{\text{RWL}}$$

		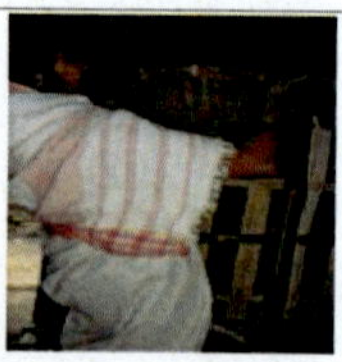
Loading paddy gains	Carrying paddy grains	Unloading paddy grains to the storage place

3 Results and Discussion

Personal and demographic characteristics of farm women revealed that cent percent of respondents were literate, and 66% of respondents belonged to nuclear families. The majority of the respondents (82%) belonged to marginal farmers having 1 acre of land for paddy cultivation. As regards to age of the respondents, 88% falls in the age group of 30–40 years

4 Determination of Lifting Index (LI) of Farm Women in Storage of Paddy Grains by Using Conventional Basket

Determination of task variables is the first step in paddy storage activity, which were used to calculate the recommended weight limit (RWL) and lifting index (LI).

Task variables:

In paddy storage activity, task variables were used to calculate the recommended weight limit (RWL) and lifting index (LI), which was the first step, and are presented in Table 1. The average horizontal location of the hands of the users in the paddy storage activity was found to be 44.73 cm. The horizontal location is the distance from the ankles of the users to the point projected on the floor directly below the middle point of the hands grasping the basket. The values range from 35 to 53 cm. The average vertical location of the respondent farm women was found to be 79.56 cm. The maximum value was found to be 93 cm, and minimum value was 63 cm. The average vertical travel distance of load lifted (paddy carrying basket) was found to be 89.26 cm. The maximum value of vertical travel distance was found to be 140 cm, and minimum value was found to be 38 cm. The difference was more in vertical travel distance due to the different height of the storage structure in farm households. Asymmetric angle (a) is the amount of degree to which the body is required to change during the storage activity. The angle of asymmetry is the amount of trunk and shoulders rotation of the farm women in degrees. The angle of asymmetry of farm women in paddy storage activity was found to be 90°. Coupling (c) signifies types of handle used for grasping the filled-up basket for storage or coupling (c) signifies types of grip where hand can comfortably wrap or grasp the handles of the basket. A good coupling or comfortable grip is one in which the hands are able to grasp the handles easily.

An optimal design container with no handles or cut-outs or irregular objects where hand can be fixed about 90^0 is termed as fair coupling. A container of less than optimal design with no handles or cut-outs or irregular objects that are hard to handle or bulky is termed as poor coupling. The coupling of quality of the conventional bamboo basket used by farm women for paddy storage activity was observed as poor (3). Poor coupling would generally require higher maximum grasp forces and increase the risk of injury. The lifting frequency of storage of paddy grains was calculated by using the average number of lifts for 15 min, and it was revealed that the average frequency was 0.72 lift/min. The average weight of the filled-up basket was found to be 14.03 kg. The average lifting duration was found to be moderate duration.

Multipliers:

Determination of six multipliers values of paddy storage activity (lifting task) for determining the recommended weight limit (RWL) and lifting index (LI) is the second step and is presented in Table 2. The average calculated horizontal multiplier value of paddy storage activity by the farm women was found to be 0.58. The average vertical multiplier value of paddy storage activity by the farm women was found to be 0.98.

Table 1 NIOSH lifting variables of the respondents in the use of conventional basket

	NIOSH lifting variables (in Metric)								
Lifting tasks	H Horizontal location (25–63 cm)	V Vertical location (0–175 cm)	D Travel distance (25–175 cm)	A Angle of asymmetry (0–135^0)	C Coupling (1 = good, 2 = fair, 3 = poor)	F Frequency (0.2–15 lifts/min)	L Avg. load lifted (kg)	Maximum load lifted (lbs.)	Duration (1, 2, 8 h)
Total	44.73	79.56	89.26	90°	3	0.72	14.03	18	1–2

Table 2 Multipliers of the respondents in paddy storage activity by using conventional basket

	Multipliers (in metric)					
Lifting tasks-paddy storage	HM = horizontal multiplier = (25/H)	VM = vertical multiplier = [1 − (0.003 |V − (75|)]	DM = distance multiplier = [0.82 + (4.5/D)]	AM = asymmetric multiplier = [1 − (0.0032A)]	FM = frequency multiplier (from Table 5 in Appendix A)	CM = coupling multiplier (from Table 7 in Appendix A)
Total	0.58	0.98	0.88	0.71	0.90	0.90

The average value of distance multipliers of storage activity was 0.88. As regard to asymmetric multiplier (AM), it was apparent that the average value of asymmetric multipliers of storage activity was 0.71. The average value of frequency multipliers of storage activity by the farm women was found 0.90. The average value of coupling multipliers of storage activity by the farm women was found to be 0.90.

Recommended weight limit (RWL):

Recommended weight limit (RWL) of farm women in paddy storage activity was found to be 6.54 kg with the use of conventional basket (Table 3 and Fig. 1). Hence, it can be recommended that the maximum acceptable weight (load) for healthy farm women should be 6.54 kg without any risk of musculoskeletal disorders (MSDs) in the lower back. Surprisingly, in the present study, respondents used to lift an average of 14.03 kg, which is beyond their permissible limits.

Lifting index (LI):

Table 3 RWL and LI of the farm women in the use of conventional basket

Sl. No	RWL = (23) (HM) (VM) (DM) (AM) (FM) (CM)	RWL	Avg. load lifted (kg.)	Lifting index (LI) = weight lifted/RWL
Total		**6.54**	**14.03**	**2.14**

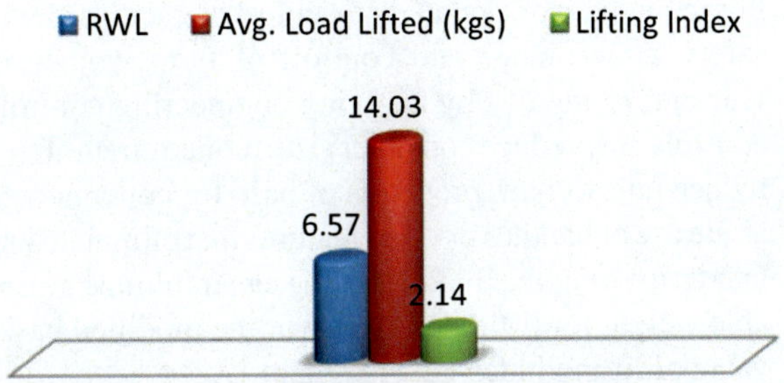

Fig. 1 RWL, average load lifted (kg), and lifting index of conventional basket

The final step of NIOSH lifting equation is the determination of lifting index (LI), which gives a relative indication of the risk of injury of respondents in paddy storage activity. The lifting index (LI) of farm women in paddy storage activity was found to be 2.14 (Table 3 and Fig. 1), which indicates that farm women suffer from a lot of work-related musculoskeletal disorders (WMSDs) or injuries. Design modification of conventional basket is necessary for reducing the risk of injury, health hazards, and increasing workable life of farm women in paddy storage activity. The result indicates that storage of paddy grains by conventional basket is highly stressful as lifting index (LI) calculated was more than 1. Similar finding were observed by Chattopadhyay et al. (2009), who conducted a study on female construction labors in West Bengal by using revised NIOSH lifting equation, and found that the weight of the materials lifted by male as well as female laborers were much more than RWL and lifting index calculated in all the lifting tasks were much more than 1. It was also found that lifting of materials above RWL produced backward bending of the back during lifting (sometimes more than 20° of extension of the back) and imposed maximum postural stress on the low back and increased the risk of lifting related low back pain and disorder in all lifting activities carried out by male and female construction labors.

5 Conclusion

The primary product of the NIOSH lifting equation is the recommended weight limit (RWL) or permissible weight limit, which defines the maximum acceptable weight (load) for all healthy workers. Determination of maximum acceptable weight can be an appropriate base for planning and implementing interventional ergonomics programs. The average weight of the load lifted by farm women was found to be 14.03 kg. The recommended weight limit (RWL) for storage activity by the use of conventional basket was found to be 6.54 kg. The lifting index of storage of paddy grains by farm women was found to be 2.14, which indicates that farm women suffer from a lot of work-related musculoskeletal disorders (WMSDs) or injury, and design modification is necessary for reducing health hazards and work-related musculoskeletal disorders (WMSDs) of farm women in storage of paddy grains. For reducing work-related musculoskeletal disorders (WMSDs), enhance efficiency, safety, convenience and comfort of farm women conventional basket should be redesign or modify by applying engineering controls. Applications of engineering controls may reduce container size, reduce circumferences and depth of the container, reduce unit weight, provision of base for container stability and grip/couplings such as the use of handles or other features that eliminate hand grip discomfort and increase handgrip strength. Good coupling can minimize grasp forces and increase the acceptable weight for lifting. Capacity of the modified basket should be able to reduce the value of lifting index (LI) less than 1.

References

Chattopadhyay S, Basu K, Sahu S, Paul G (2009) Ergonomic evaluation of postural stress of male and female construction laborers' employed in unorganized sectors in West Bengals. In: Gangupadhyay S (ed) Ergonomics for everyone. Department of Physiology, Occupational Ergonomic Laboratory, University of Calcutta, pp 131–135

Das T, Gangopadhyay S (2005) A critical evaluation of the probable causes of musculoskeletal disorders prevalent among the manual material handling workers of a construction. In: Proceedings of international ergonomics conference on humanizing work and work environment. IIT, Guwahati, pp 283–287

Gangopadhyay S, Das T, Ghoshal G, Das B (2003) An ergonomic study on the development of musculoskeletal disorders among the fixed load and variable load handling workers of Kolkata. Indian J Biol Sci 9:37–45

Nelson GS, Wickes H, English JT (1994, 2010) Manual lifting: the NIOSH work practices guide for manual lifting determining acceptable weights of lift. Nelson and Associates

Chapter 5
Ergonomic Design of Vegetable and Fruit Cutter

Udakam Vamsi Krishna, Potturi Phani Kumar, Buddana Divya Sri, and S. A. Basha

1 Introduction

Nowadays, every field has embraced with automation, from manufacturing to food processing, biomedical, and pharmaceutical industries. Of late, processes that were manual before are slowly being converted to semi-automated and automated. In such a scenario, food preparation-related activities also got attention by researchers to automation to a possible extent. Most of the vegetables such as potato, onion, cucumber, as well as fruits like banana, papaya are served in slices for making the food more delectable. Cutting of vegetables manually is still prevalent, in food courts, catering services, and restaurants, which can cater food to a whole set of varying customer tastes and preferences. The amounts of vegetables to be cut are associated with difficulties like time consumption, contamination, injuries etc. make it pretty difficult for a person handling the job, thereby arose a need to simplify the processes of the vegetable and fruit cutting as per the user requirement.

In this regard, many researchers developed machines based on various mechanisms to cater the need. They have been presented in the following section along with the features of the proposed machine.

2 Literature Survey

The design proposed by Tawi (2014) was based on crank and slotted lever mechanism, which seems to be less compact but more complex. Moreover, the mechanism was driven by manually. Ankit and Patra (2015) developed a machine based on

U. V. Krishna · P. P. Kumar · B. D. Sri · S. A. Basha (✉)
Department of Mechanical Engineering, Rajiv Gandhi University of Knowledge Technologies, Nuzvid, Andhra Pradesh 521202, India
e-mail: n130065@rguktn.ac.in; basha@rguktn.ac.in

© The Author(s), under exclusive license to Springer Nature Singapore Pte Ltd. 2022, corrected publication 2022

D. Bijulal et al. (eds.), *Technology-Enabled Work-System Design*, Design Science and Innovation, https://doi.org/10.1007/978-981-16-1884-0_5

two mechanisms, i.e., Crank–Rocker mechanism and the Scott–Russell mechanism clubbed together in view to help fish sellers. It is motor driven and uses heavy links. Though feeding of the vegetables is automatic but power consumption and torque requirement are high.

Thomas et al. (2014) developed a machine, which uses a programmed Adriano and servo motor. It consists of a hopper to feed vegetables and a pneumatically controlled pressure block to control grid apparatus. A tray is placed at the bottom of the apparatus to collect the vegetable pieces after processing. Variable pressure setting for cutting different vegetables is carried out by the microcontroller. The system has several advantages; however, automation leads to high equipment cost as well as power consumption. Naveen et al. (2016) developed multi-purpose kitchen equipment that can perform the operations such as grinding rice flour, vegetable cutting, and coconut scrapping. This is based on slider crank mechanism and operated by motor. The design is complex, and equipment is heavy that limits the portability. Since the motor runs at high rpm, and total torque is transmitted to the mechanism, this may cause injuries to the operator.

From the navigation of literature survey, the drawbacks identified in the existing machine are high investment cost, design complexity, contamination, safety of the operators, and high-power requirement. In this regard, we proposed a slider crank mechanism-based machine, which is operated manually, has safeguards, economical and portable. During design process, the ergonomic principles and anthropometric measurements were considered. The detailed description of design of machine elements, mechanism, and analytical analysis is presented in the following section.

3 Analytical Calculations

The main components of slider crank mechanism are crankshaft, connecting rod, and piston. The connecting rod converts the rotary motion of crank into reciprocatory motion, which, in turn, slides piston to and fro. The materials of crankshaft are mild steel, connecting rod is cast iron, and piston is forged steel. A typical slider crank mechanism is shown in Fig. 1 [Courtesy (Rattan 2017)], in which the stroke line of

Fig. 1 Slider Crank mechanism

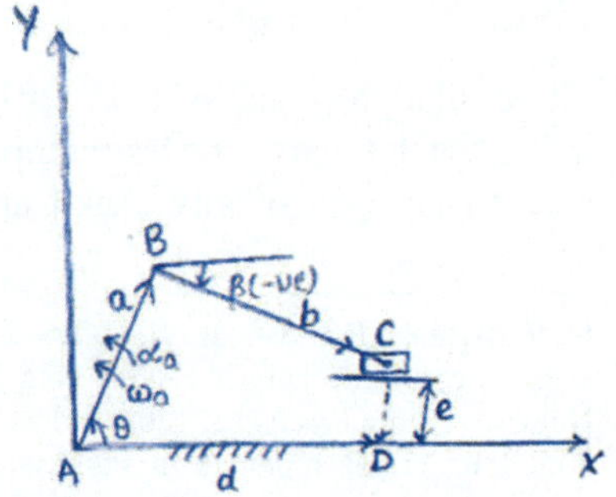

the slider (piston) does not pass through the axis of rotation of the crank. Angle β is the angle made between slider and connecting rod. Angle β in clockwise direction from the x-axis is taken as negative. Angle θ is the angle made between the crank and base. Crank is rotating with angular velocity ω_a. The relationship between various parameters of mechanism is presented as follows.

Let e = eccentricity (distance CD) = 0.

AB = a = Crank Length (or) Crank Radius.

BC = b = Length of Connecting rod.

Displacement along x-axis,

$$a\cos\theta + b\cos(-\beta) = d$$

$$b\cos(\beta) = d - a\cos\theta \tag{1}$$

Displacement along y-axis,

$$a\sin\theta + b\sin(-\beta) = e$$

$$b\sin(\beta) = e - a\sin\theta \tag{2}$$

Squaring Eqs. (1) and (2) and adding,

$$b^2 = a^2\cos^2\theta + d^2 - 2ad\cos\theta + a^2\sin^2\theta + e^2 - 2ae\sin\theta$$

$$= a^2 + e^2 + d^2 - 2ae\sin\theta - 2ad\cos\theta$$

or

$$d^2 + a^2 - b^2 - (2a\cos\theta)d + e^2 - 2ae\sin\theta = 0$$

or

$$d^2 + C_1 d + C_2 = 0 \tag{3}$$

where $C_1 = -2a\cos\theta$

$$C_2 = a^2 + e^2 - b^2 - 2ae\sin\theta$$

Equation (3) is a quadric in d, its two roots are,

$$d = \frac{-C_1 \pm \sqrt{{C_1}^2 - 4C_2}}{2}$$

Thus, if the parameters a, b, e and θ of the mechanism are known, the output displacement can be computed. Also from Eq. (2)

$$\beta = \sin^{-1}\frac{e - a\,\sin\theta}{b}$$

3.1 Calculation of Diameter of Piston and Crankshaft

From the literature, crank angle (θ) is taken as 60° (Naveen et al. 2016), which results in β = 30°. The crank length or crank radius (a) is assumed as 3.5 cm, then stroke length (l) should be double of the crank radius (a) i.e., $l = 2a = 2*3.5 = 7$ cm.

Piston diameter = 7 cm.

Angular velocity of Crank (ω_a) = 2.47 rad/s.

Rotations per minute (N) = $\frac{\omega_a*60}{2\pi}$ = 24 rpm.

Linear velocity of piston = $\frac{a*\omega_a*\sin(\beta-\theta)}{\cos\beta}$ = 5 cm/s.

3.2 Calculations of Connecting Rod and Time Analysis

From Eq. 2, connecting rod length (b)

Connecting rod length (b) $\Longmapsto b\sin(\beta) = e - a\sin\theta$

$\Longmapsto b = \frac{e-asin\theta}{\sin\beta}$

= 6.06 cm.

Equation of freely falling body is

$$S = ut + \frac{1}{2}gt^2$$

where,

S = Displacement or stroke length.

u = initial velocity of the particle.

t = initial time or starting time.

g = gravitational constant = 9.8 m/s^2

Here initial velocity u = 0 so.

ut = 0

$$S = \frac{1}{2}gt^2$$

$7 = \frac{1}{2}9.8t^2 \Longmapsto t^2 = \frac{7*2}{9.8} = 10/7 \Longmapsto$ t = 1.19 s.

Velocity of piston = 5 cm/s.

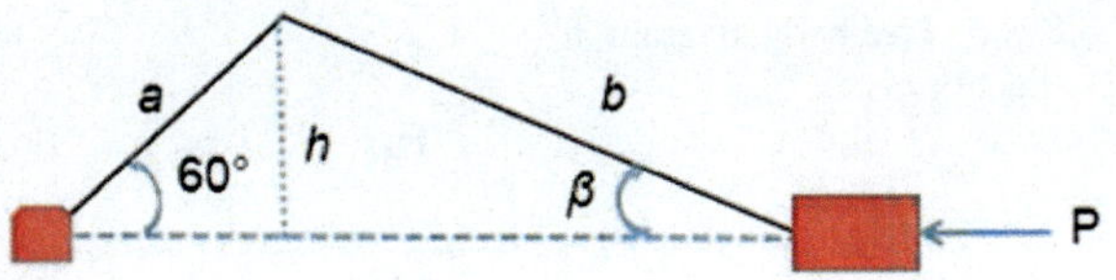

Fig. 2 Static force analysis for slider crank mechanism

Displacement of piston or stroke length = 7 cm.

Piston time to move from Bottom Dead Centre [B.D.C] to Top Dead Centre [T.D.C] is

$$velocity = \frac{displacement}{time} = \frac{7}{time}$$

$$\text{Time} = 1.4 \text{ s}$$

Time = 1.4 s

3.3 *Force Analysis*

The calculations related to force required to cut the vegetable without crushing are presented as follows [Bhandari (2017)]. The freebody diagram is shown in Fig. 2.

Let, P = resisting force offered by
vegetable for potato, $P = 100$ N [7]
θ = crank angle = 60°
β = angle made by the slider with connecting rod = 30°
a = crank radius = 3.5 cm = 35 mm
b = connecting rod length = 14 cm = 140 mm
Perpendicular height, $h = r \sin\theta = l \sin\beta$
h = 3.5 *Sin (60°) = 3.03 cm

3.3.1 Forces Acting on the Slider

The freebody diagrams of forces acting on slider are shown in Fig. 3. $F_{14}{}^{y}$ is the force acting on the slider by the action of fixed link. F_{34} is the force acting on the slider by the action of the connecting rod. $F_{34}{}^{x}$, $F_{34}{}^{y}$ are the resolutions of the forces in the direction of x-axis and y-axis.

$$\tan\Phi = \frac{F_{34^y}}{P}$$

$$F_{14^x} = 0$$

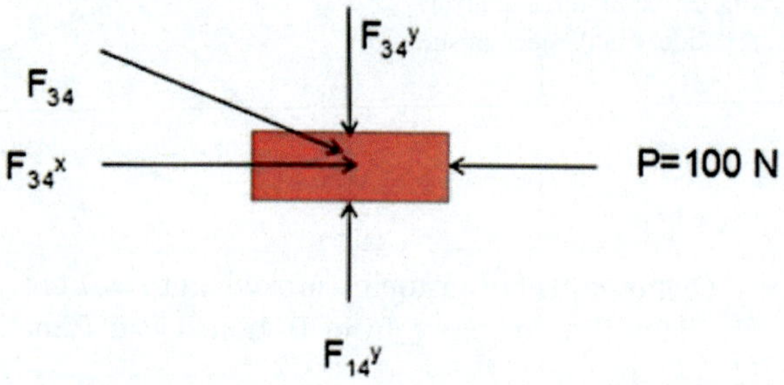

Fig. 3 Free body diagram of slider or piston

$$F_{34^y} = -P \tan \Phi$$

$$F_{14^y} = P \tan \Phi = 57.735\,\mathrm{N}$$

$$F_{34^x} = \mathrm{P} = 100\,\mathrm{N}$$

3.3.2 Forces Acting on the Connecting Rod

The free body diagram of forces acting on connecting rod is shown in Fig. 4. $F_{23}$is the force acting on the connecting rod by the action of crank. $F_{23}x$ and $F_{23}y$ are the resolutions of the forces in the x and y directions, respectively. F_{43} is the force acting on the connecting rod from the slider and then resolved into $F_{43}x$, $F_{43}y$ in the direction of x and y, respectively.

$$F_{43^x} = -\mathrm{P} = -100\,\mathrm{N}$$

$$F_{43^y} = P \tan \Phi = 57.375\,\mathrm{N}$$

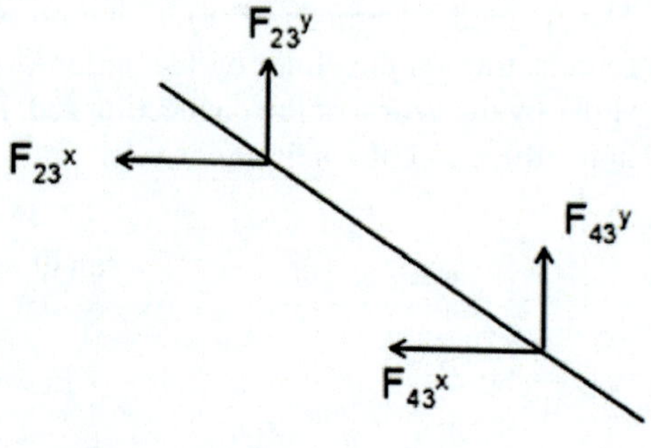

Fig. 4 Free body diagram connecting rod

Fig. 5 Free body diagram of Crank

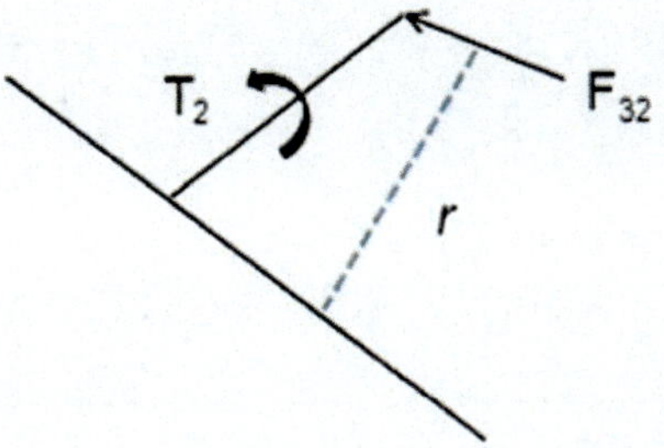

Fig. 6 Angle made by crank

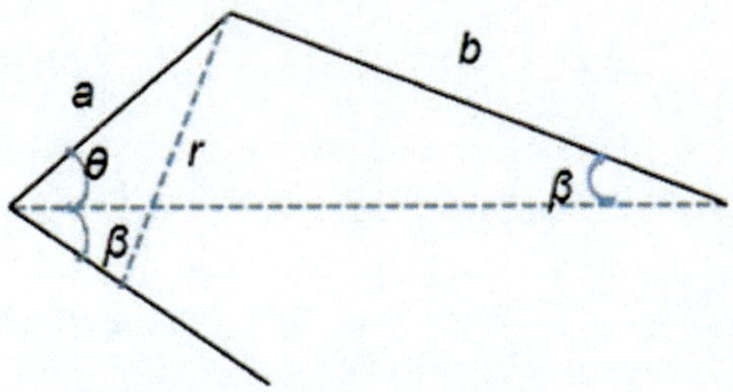

$$F_{23^x} = P = 100\,\text{N}$$

$$F_{23^y} = -P\tan\Phi = -57.375\,\text{N}.$$

3.3.3 Forces and Torque Acting on the Crankshaft

The free body diagram of forces and torque acting on the crank is shown in Fig. 5. T_2 is the torque acting on the crank. F_{32} is the force acting on the crank from the reaction of force of connecting rod. r is the perpendicular distance for F_{32} (Fig. 6).

$$F_{32^x} = -\text{P} = -100\,\text{N}$$

$$F_{32^y} = \text{P}\tan\Phi = 57.375\,\text{N}$$

$$T_2 = F_{32} * \text{d}$$

$$\text{Sin}(\theta + \Phi) = \frac{r}{a}$$

$$\theta = 60;\ \Phi = 30$$

$$\mathrm{Sin}(60+30)=\frac{r}{a}$$

$$r=a\sin(90)$$

$$\mathrm{r}=3.5*1$$

$$\mathrm{r}=3.5\ \mathrm{cm}$$

$$T_2=F_{32}*\mathrm{r}$$

$$F_{32}=\frac{P}{\mathrm{Cos}(\Phi)}$$

$$T_2=P*\mathrm{Sec}(\Phi)*\mathrm{r}$$

$$=100*(3.5*10^{-2})*\mathrm{Sec}(30)*\mathrm{Sin}(90)$$

$$=4.04\mathrm{Nm}$$

Torque required to rotate the Crankshaft = 4.04 Nm.

The proposed machine consists of a guiding cylinder, hopper block, and blade set. The slider reciprocates to (TDC) and fro (BDC) in the guiding cylinder and applies force on the vegetables/fruit. The hopper block is used to feed the vegetables/fruit. The blade set is arranged on the other end (TDC) of the guiding cylinder.

From the above calculations, it is observed that the force and torque required to cut the vegetable are within the limits of hand-driven force. The time estimates for vegetable to settle in the chamber and the slider to move from BDC to TDC are 1.19 s and 1.4 s, respectively. This enables the slider to slice the vegetable present in the cylinder instead of crushing it.

In the following section, (i) the CAD model developed by using Solid Works Software, and (ii) the prototype developed is presented.

4 Prototype

According to the dimensions obtained from the design of machine elements, a CAD model of the proposed machine is developed using Solid Works software. Initially, individual components were drafted, then assembly of these parts is carried. The typical Solid Works model of assembled proposed machine with specified names is shown in Fig. 7.

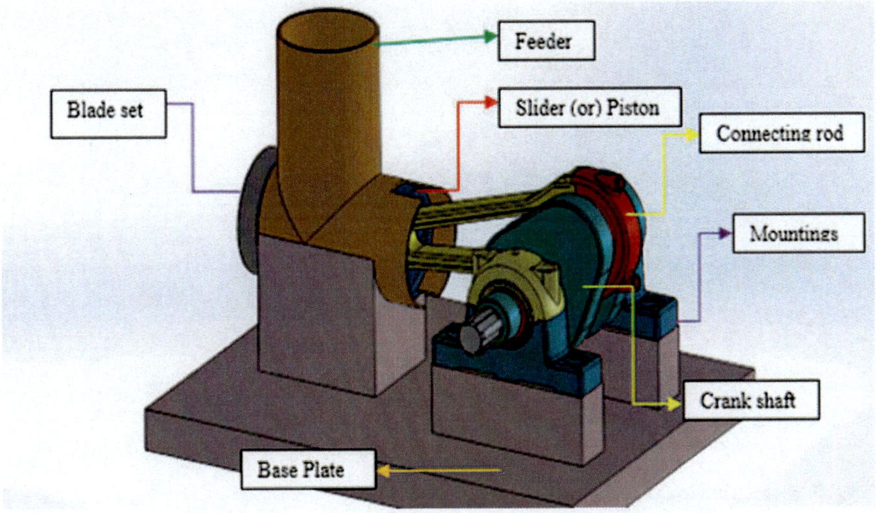

Fig. 7 Solid works model

Provision is given to accommodate a removable blade set at the end of the guiding cylinder. This allows the user to obtain a variety of shapes and sizes of pieces of vegetables/fruits. A typical CAD model of blade set is shown in Fig. 8.

After assembly, simulation motion analysis is carried to check the interference between rotating parts and moving parts. From the simulation study, it is confirmed that there is no interference between mating parts. Finally, a real-time prototype of the proposed model is developed as per the dimensions obtained from the design calculations.

The steps involved in prototype development are (i) a wooden base is prepared to mount all the components on it, (ii) the crank and connecting rod assembly are fitted between two foot bearings, (iii) a handle is attached to the crank shaft, (iv) the piston is arranged in the guiding cylinder, (v) the connecting rod is attached to piston, (vi) blade set is fixed at the end of guiding cylinder. The top view and the side view of prototype developed are shown in Fig. 9 and Fig. 10, respectively.

Fig. 8 Blade set

Fig. 9 Isometric view of the prototype

Fig. 10 Prototype isometric view

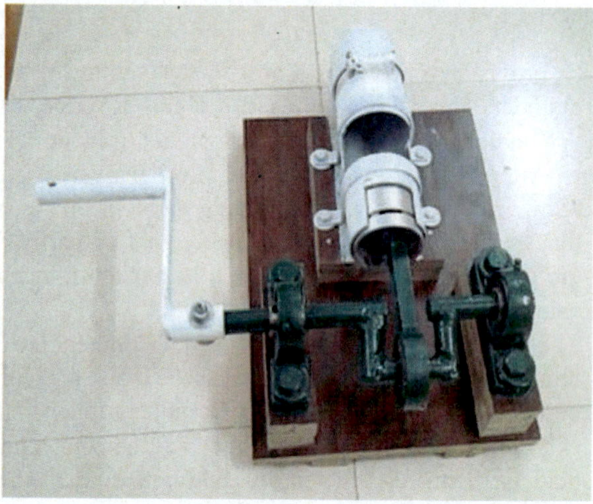

5 Conclusions

Most of the existing models are power-driven, complex design, high cost, and prone to injuries to operators. In contrast, the proposed machine is manually operated, simple design, low cost, and less chance of injuries to operators. The design is tested using simulation motion analysis, and a real-time prototype is developed. This machine has the provision to change the blade set. This machine will reduce the labor hours and effort, significantly. The machine is light in weight and portable. This design has limitations with respect to the size and shape of the vegetable/fruits such as (i) can accommodate approximate circular shapes and (ii) large size and lengthy one can't accommodate. However, with proper modification in design, the above limitations can overcome.

References

Ankit P, Patra MR (2015) Design and fabrication of automatic cutting machine. B. Tech project. National Institute of Technology Rourkela

Bhandari VB (2017) Design of machine elements, 4th edn. Tata McGraw Hill Education Private limited, New Delhi

Naveen J, Manikandan A, Maheshkumar P, Selvaraj M (2016) Design and fabrication of low cost multi-purpose kitchen equipment. Int J Eng Res 4(3):163–169

Rattan SS (2017) Theory of machines, 4th edn. Tata McGraw Hill Education Public limited, New Delhi

Tawi JDA (2014) Industrial design and manufacturing of manual driven low-cost banana slicing machine

Thomas T, Muthukrishnan A, Sre Nandha Guhan KS (2014) Design and development of automated vegetable cutting machine. Design Res Conf (AIMTDR 2014) 651

www.Solidworks.com

Chapter 6
Modeling of Work-Related Health Disorders Among Women Agricultural Workers

Kiran Mohan, V. Madhusudanan Pillai, S. Abhinav, Amal Sunny, Vishnu V. Kumar, O. R. Rohith Raj, S. Abishek, and Vahid Mohammad

1 Introduction

According to World Health Organization (World Health Organization, 2004), "an occupational diseases, or work-related health disorders (WRHDs), are any diseases as a result of an exposure to risk factors arising from work activeness. Work-Related Diseases (WRDs) have multiple causes, where agents in the work environs may play a role, together with other risk factors, in the development of such diseases". Among WRHDs the most common disorders are musculoskeletal disorders (MSDs). These disorders result in permanent/temporary damages to the nerves, muscles, bones, tendons and ligaments. The development of WRHDs results in huge treatment expenditures and loss of productivity and hence loss of job. The cause of work-related diseases can be due to several factors, which can be detectable or hidden, solely occupational or a combination with non-occupational factors. The symptoms of these disorders are noticeable from 3 to 5 years of work. The development of WRHDs among the working class is a growing matter of concern.

According to International Labor Organization (International Labor Organization, 2001), worldwide, an estimated 1.3 billion workers are working in agricultural sectors. This forms half of the labor force of the world. Most dangerous occupational categories causing work-related disorders are construction, mining and agricultural sector, and yearly at least 1,70,000 workers are killed in agricultural sector. That is, compared to other sectors, workers in agriculture sector have twice the risk of dying on the job" (International Labour Organization). According to World Bank Report (World Bank Report), more than 41% of people are employed in agriculture and related activities in India, among them, 56% are women workers. Agricultural

K. Mohan (✉) · V. M. Pillai · S. Abhinav · A. Sunny · V. V. Kumar · O. R. R. Raj · S. Abishek · V. Mohammad
Department of Mechanical Engineering, NIT Calicut, Calicut, India

V. M. Pillai
e-mail: vmp@nitc.ac.in

© The Author(s), under exclusive license to Springer Nature Singapore Pte Ltd. 2022, corrected publication 2022
D. Bijulal et al. (eds.), *Technology-Enabled Work-System Design*, Design Science and Innovation, https://doi.org/10.1007/978-981-16-1884-0_6

activities require huge physical demand; workers face the ill effects of it. As per ILO report (International Labour Organization), "the risk of accidents is increased by fatigue, poorly designed hand tools, difficult terrain, exposure to extreme weather conditions, and poor general health, associated with working and living in remote and rural communities". As far as women workers are concerned, they have to work in the field and as well as take care of household activities. Moreover, they are considered to be physically weak than men. Through this study, it is intended to find out the triggering factors leading to the development of WRHDs on the basis of which measures can be taken to reduce or eliminate the effect of these factors, hence improving the working condition of women agricultural workers. The various objectives of the study are:

- To find the predominant work-related disorders among women agriculture workers.
- To verify whether there is a significant difference in disorders existing in the agriculture workers of different locations in Kerala.
- To find the various factors and their level of contribution.
- To assess workplace based on factors that affect occupational health.
- To develop a model using Structural Equation Modeling techniques.

2 Literature Review

Agriculture is a field where workers are required to physically work hard under extreme conditions, and hence, the development of WRHDs is very likely among agricultural workers. According to Fathallah (2010), agricultural workers are involved in labor-intensive practices, and hence, they are exposed to a multitude of MSDs risk factors. Various factors influencing WRDs in different occupations were found using the review of literature, and the results obtained were used for the preparation of the data collection tool. A study conducted by Meyers et al. (2008) says that the three primary risk factors in agriculture are awkward postures, lifting and carrying heavy loads and repetitive handwork. According to Das et al. (2012) and Das (2011), the common disorder among agricultural workers is pain caused in lower back region, shoulder, knee, ankles, feet, hands, etc. among which the most predominant one is the lower back pain. This is supported by the findings of Walker-Bone and Palmer (2002) and Xiang et al. (1999).

Apart from MSDs, there are other disorders caused due to work like respiratory, skin, eye, ear, allergic problems and so on. Borah (2015) found that the release of toxic smoke and suspended particles into air lead to respiratory problem, asthma and eye irritation. A study conducted by Stanbury et al. (2008) indicates that noise in their working environment may lead to deafness or trouble in hearing of one or both ears. A positive association is identified between smoking and back pain in a study conducted among the general population by Palmer et al. (2003). Along with physical factors, psychological factors like workload, time pressure, monotonous work, etc. can also cause WRHDs. This relationship between physical and psychological factors toward the development of disorders has been proven in a study conducted by Devereux et al. (2002). This study found a positive correlation between exposure to both physical and psychosocial risk factors at work and symptoms of disorders; supported by the study performed by Parkes et al. (2005).

One of the causes of WRHDs is repetitive work without proper rest. Lim and Kong (2014) found that the average reduction of total grip strength is increased with decreasing resting time. There is an inverse relationship between reported disorders and leisure time according to a study by Halldin et al. (2007). Education, socioeconomic factors and lifestyle can also cause disorders. This is supported by a study conducted in France by Yarnell et al. (2005). Women agricultural workers are more likely to be affected by the WRHDs as most of the tools they use are developed based on male dimensions. Moreover, women, in general, are physically weaker. The studies conducted on the prevalence of WRHDs among women in India and especially in the southern part are limited. Since women are engaged in both occupational and household activities, insufficient rest time and mental stress add to the development of WRHDs among women.

Literature suggested several triggering factors causing WRHDs and main factors are working posture, repeated activity, work environment, demography, work–life balance, socio-economic status. All of these factors will not be the triggering factor in every occupation. The factors and their degree of contribution toward the development of WRHDs depend on the type of work. For example, work environment may not affect a video display terminal (Ranasinghe et al. 2011) worker who spent most of his time in an air-conditioned room. He is likely to be affected by the design of the chair. However, the working environment has a greater influence on a farmer working in open or person working in a mine. Women agricultural workers are likely to be affected by all the factors in the above-mentioned groups. Working posture plays a major role in developing WRHDs in any occupation. Being in an awkward posture for a prolonged time can strain the muscles causing pain (Shojaei et al. 2016). Similarly, bending (Singh et al. 2007; Ciriello and Snook 1999), twisting (Bengal et al. 2005; Rosecrance et al. 2006; Basahel 2015) and lifting heavy loads (Parikh et al. 1997; Lee and Chen 1996; Song and Qu 2014; Dempsey 2003) can cause the dislocation of joints and pain in the back region. Carrying heavy loads and overhead works (Tanaka et al. 2001; Basahel 2016) are detrimental in causing neck pain. Repetitive work without sufficient rest can result in the straining of certain region ultimately causing pain in those regions. Apart from occupational factors, there can be the contribution of non-occupational factors in the growth of WRHDs. Several studies have identified the working posture-related discomfort among workers. They are improper seating (Mehta et al. 2008; Dhimmar et al. 2011; Wang et al. 2004; Rahmatalla and Deshaw 2011; Tan et al. 2008), twisting and bending (Singh et al. 2007; Ciriello and Snook 1999; Rosecrance et al. 2006; Das et al. 2013; Agbor and Hilbert 2014; Choobineh et al. 2006), prolonged standing (Osborne et al. 2012a, 2010; Burdorf and Sorock 1997; Burdorf et al. 1991) and lifting (Parikh et al. 1997; Lee and Chen 1996; Song and Qu 2014; Dempsey 2003; Jager and Luttmann 1999; Beek et al. 2005; Nimbarte 2014; Visser et al. 2014).

Experience has an inverse as well as the direct effect in causing disorders. Either an experienced worker has more possibility of receiving WRHDs as he has been exposed to extreme conditions for a longer period or he is less likely to have WRHDs as with experience he comes to know about the ill effects of doing work improperly, so he will do the work in the right way avoiding the ill effects. Rest (Osborne et al. 2012b) and leisure activities (Tai and Tam 1997) not only help the worker to get rid of the fatigue but also reduce mental stress, which can also cause WRHDs. Improper sleep and diet result in the weakening of body, which worsens further when exposed to extreme conditions. Lifestyle also contributes to causing WRHDs. Many studies have pointed out that the consumption of alcohol, use of tobacco, smoking (Waghorn et al. 2006) can affect the nervous system gifting WRHDs at last.

Several factors associated with working environment contribute to disorders. Some of them are occupational allergens (Niu 2000; Mobed et al. 1987), exposure to hazardous environment (Majumder 2014; Litchfield 1999), transmitted vibration (Tan et al. 2008; Xu et al. 2016; Dewangan and Tewari 2010; Azmir et al. 2015; Cherian et al. 1996; Adewusi et al. 2013; Albers et al. 2005; Mandal and Srivastava 2006; Gupta 2013) and occupational stress (Moon and Sauter 2005). The extreme conditions in the working environment easily trigger disorders. The dust, plants and pollen grains can cause various allergic problems (Mobed et al. 1987; Bahri 2014; Gangopadhyay et al. 2015; Srinivas and Ramanathan 2005; Kolstrup 2012). The use of powered saws and similar tools causes vibration-induced disorders (Tan et al. 2008; Xu et al. 2016; Dewangan and Tewari 2010; Azmir et al. 2015; Cherian et al. 1996; Adewusi et al. 2013; Albers et al. 2005; Mandal and Srivastava 2006; Gupta 2013). High temperature and humidity cause dehydration and problems like sunburn. Inhalation of smokes and poisonous fumes results in respiratory problems like work-related asthma. Socioeconomic factors also influence causing WRHDs. Many studies show that illiterate workers are more prone to have disorders. Most of the working-class people have poor financial backgrounds (Salas et al. 2016). Moreover, in order to earn, they often work in extreme conditions beyond their normal working hours and sometimes during night-time in poor illumination. As pointed out by many studies, women suffer the ill effects of WRHDs than men. There can be several explanations for this. Most of the tools and working environment are designed based on male ergonomics and hence uncomfortable for women workers. Various body dimensions (Mehta et al. 2008) play a role in developing WRHDs. For example, a tall worker will find it difficult to use a spade with a short handle as he has to bend more and vice versa. With age (Srinavin and Mohamed 2003), body weakens, and hence chances of getting disorders are high.

The present review indicated that during subjective evaluation of Work-related Disorders (WRDs), prevalence and occurrence of WRDs and MSDs mainly depend on the nature of work and working conditions. Studies coined above reveal that the effect due to factors such as demography, work-life balance, job satisfaction, working posture, repeated activity, exposure to hazardous working conditions, socioeconomic status, habits, etc. should be given ample importance as these factors are the major contributors toward WRDs. Articles related to extend of these factors in contributing

WRDs are limited. Extend of these factors can be identified by developing a mathematical predictive model considering these factors as independent variables. A mathematical model for finding the extent of the contributing factors is discussed below.

2.1 Proposed Model

From review of literature, it is understood that contributors of work-related disorders are mainly the contributors of MSDs; and in most cases, other triggering factors causing disorders are not considered. Therefore, to compensate the limitation, a new model is proposed, which is a combination of various models described in different literature. In the proposed model, work-related disorders are expressed as a function of triggering factors or contributors. Factors contributing MSDs as well as other disorders such as work environment, demography, work-life balance, habits and socio-economic status are also considered for the expression.

Work-related disorders = Function (Working posture, Work environment, Demography, Work–life balance, Habits and Socio-economic status)

where,

- Working posture is considered as a function of improper seating, overhead works, bending and twisting, prolonged standing or lifting
- Work-life balance is considered as a function of experience, rest time, sleep, diet, leisure activities or job satisfaction
- Demography is a function of temperature, humidity, height and weight, body dimensions or age
- Socio-economic status is a function of marital status, education or economic status
- Work environment is a function of occupational allergies, vibration, stress/strain or exposure to hazardous environment
- Habits is a function of alcohol, use of tobacco, fast food, skipping meals or anxiety

Using the findings from reading the literature and taking reference to the standard Nordic questionnaire, the study has developed a schedule for data collection. Details related to the survey instrument (schedule) are given in Sect. 3.1, and different validity measures and reliability of the schedule are also provided in this section.

3 Method

3.1 Survey Instrument

The triggering factors are identified by referring to literature on agriculture in India, and factors are categorized based on their source and its effect. The primary tool used for data collection is a schedule. The schedule consisting of questions related to WRHDs is prepared by taking reference of standard Nordic questionnaire and new questions regarding agriculture-based disorders are added by referring literature in this field. The questions consist of both open-ended and closed-ended pattern. The reason for choosing a schedule is that it consists of a formalized set of questions and statements with space for answers. The data collectors administer the questions to potential respondent and note down the response. The schedule consists of the following sections.

- Demographic details: Various demographic factors such as age, height, weight, education, etc.
- Work characteristics: number of working day, hours spent in a day, etc.
- Workplace assessment: details regarding various factors of the work environment.
- Repeated activity: details regarding rest time, nature of the work.
- Lifestyle: various lifestyle factors like consumption of junk food, use of tobacco, etc.
- Musculoskeletal discomfort: pain in various body regions.
- Post-injury survey: details related to injury, treatment, loss of work, etc.

Different validity measures such as linguistic, content and criterion are done based on rational analysis by raters (or experts in the research subject). Construct and discriminant validity is done after the study. The feasibility of the approach is examined using a pre-test of the research instrument (schedule). In the pre-test, the questionnaire is tested in order to identify any complications related to unclear wording or response and also to know an estimate of the duration to administer the schedule. In this study, 30 samples are collected and from the responses obtained some modifications are done in the schedule for getting more accurate responses. It is possible to reproduce the result under a similar study and hence the research instrument is reliable. The reliability is done by following Cronbach's alpha method. The reason to choose this method is because it is the most widely used objective method for reliability (Meyers et al. 2008).

3.2 Data Collection

The objective is to find the factors triggering WRHDs among women agricultural workers. For this purpose, a sample is taken from the women agricultural workers employed under Mahatma Gandhi National Rural Employment Guarantee Scheme (MGNREGS). According to the information from the official website of MGNREGS, there are about 20 lakhs of women working under this scheme in Kerala. So the population is assumed to be infinite for sample size calculation. The formula used for calculating the sample size is given below:

$$n = \frac{z^2 \times p(1 - p)}{e^2}$$

where,

n is the sample size

z (1.96) is the selected critical value of desired confidence level

p (0.5) is the estimated proportion of an attribute that is present in the population

e (0.05) is the desired level of precision.

Substituting values in the equation, the sample size comes out to be 385. The sampling technique used is convenience sampling. A total of 436 women workers employed under the scheme of MGNREGA were interviewed from nine panchayats for the study. 64% have the total experience above 40 years, 22% have an experience between 20 and 30 years, 10% have an experience between 10 and 20 years and 4% have an experience below 10 years. Therefore, the study is conducted among people who are having higher chances of occurrence of MSDs. In case of age, major part (40%) comes under 40–50 years followed by 50–60 years (30%), 30–40 (16%) years and above 60 years (14%). The questions were delivered in Malayalam, which is the mother tongue of all the participants. Hence, it was easy for the workers to understand the question properly and provide the right answers. It took around 7–15 min to interview one worker. The weight of the workers were measured using a weighing machine and height using a meter tape.

Places of data collection were selected based on different geographical zones (inland, hilly and coastal areas) of Kerala, and a total number of six districts were visited.

- In **Kozhikode** district, three panchayats (Ayancheri, Changaroth and Kodancheri) were covered.
- In **Wayanad** district, Vythiri panchayat was covered.
- In **Palakkad** district, Kodumbu panchayat was covered.
- In **Kannur** district, Kariyadu panchayat was covered.
- In **Thrissur** district, Madakkathra panchayat was covered.
- In **Malappuram** district, two panchayats (Vallikkunnu and Areekode) were covered.

3.3 Data Analysis

Data analysis is the sequential process leading up to the development of a mathematical model. Initially, data screening is conducted to eliminate outliers. Outliers are data points having large variance and have to be eliminated to obtain reliable results. Outliers are removed via two methods, viz., univariate analysis and multivariate analysis. Boxplots are employed as a univariate analysis technique and Mahalanobis distance as a multivariate technique, 23 samples were removed as outliers. Following this, variables are reduced by carrying out a factor analysis. There are basically two methods for factor analysis, viz., principal component analysis (PCA) and exploratory factor analysis (EFA). The difference between the two methods is that the exploratory factor analysis is used for deriving a mathematical model from which factors are estimated, whereas the data are separated into its constituent parts in PCA (Field 2013).

Exploratory factor analysis (EFA) is a statistical analysis whose goal is to identify the relationship between the observed variables. The objective is to reduce the number of variables and group them into factors having highly correlated variables. There are two types, R-Type and Q-Type. R-Type uses the correlation matrix of variables to define factors while Q-Type uses individual respondents to obtain factors. Q-Type is generally considered as it is easier to group variables based on their correlation value. For driving the factors, principal component analysis (PCA) is used. This method is used to extract minimum number of factors explaining maximum number of variance. It considers total variance which consists of unique variance, common variance and specific variance. Another method is common factor analysis, which uses the common variance, which does not consider unique and error variances. This method is used if there is no prior knowledge of the nature of the variables. Since this study focuses on obtaining minimum number of factors covering maximum variance of variables, PCA is used.

The criteria of the number of factors extracted are based on the above methods. But to know if the factor provides more adequate information about the variables examined, factor rotation must be done. Factor rotation is used to simplify the factor structure and consider the maximum distribution of variance among variables. Two methods are used mainly, Orthogonal Rotation and Oblique Rotation. Orthogonal is used when factors are considered not to have any correlation; it uses *varimax*, *quatrimax* and *equimax* methods. Oblique is considered when there is some sort off correlation considered between factors. SPSS provides *oblimin* rotation for oblique rotation. This study considers oblique rotation (*oblimin*) as there might be some correlation within the factors obtained. This categorizes the remaining variables into meaningful factors having some physical significance, which can then be appropriately concluded. A reliability analysis is carried out after this by inspecting the value obtained for Cronbach's alpha. Dependent variable is the prevalence of WMSD, which is represented as binary (0, 1) in which '1' indicates that the worker possesses WMSD and '0' indicates that the worker is free from WMSD. A person who has pain in a particular area for more than 30 days in a year is considered to have WMSD and the factor score generated for each factors are saved as new variables for further analysis. The factors are taken as independent variables to conduct logistic regression. Logistical regression method is similar to linear regression where the output is dichotomous (only two possible outcomes). This method is used if there is more than one independent variable to determine the outcome. The regression will give an equation from which we can interpret the effect of the factors in causing WMSD for any worker.

Structural equation model (SEM) is a statistical methodology that is used for confirmatory factor analysis (CFA). The purpose of SEM is to develop a structural equation of the variables taken into consideration and to pictorially model a structural relation that can provide a clarity in the conceptualization of the theory under the study (Hair et al. 2010). The process of developing SEM involves developing a hypothesized model initially, input the measured variables in the independent variable

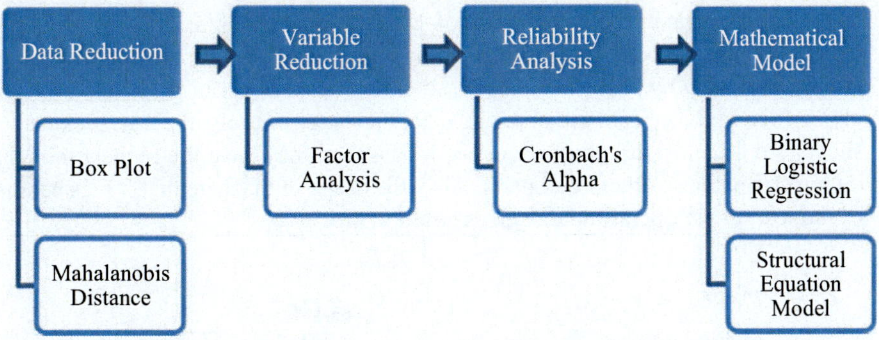

Fig. 1 Flow diagram of data analysis

box, and then run the model. On completing a successful run, coefficient of variables is obtained, and these represent how much the measured variables influence the factors.

In CFA process, the goodness of fit between the hypothesized model and the resulting model is evaluated. Also, it enables to simultaneously inspect a series of correlation among the measured variables and factors and inter-relationship between the factors (Hoefkens et al. 2012). The model fit evaluation involves in three different models: Default model, which is specified by the user. Saturated model: this contains a large number of parameters as input to the analysis. Independence model: this is the most restrictive model, which contains the variances of considered variables only, which means that it assumes zero covariance (Hamilton 2013). The indices used for test for model fit are Chi-square value, Goodness of Fit Index (GFI), Adjusted Goodness of Fit Index (AGFI), Comparative Fit Index (CFI), Normed Fit Index (NFI), Tucker-Lewis Index (TLI), and Root Mean Square Error of Approximation (RMSEA).

This study used IBM SPSS Amos for developing SEM. The flow diagram (Fig. 1) of data analysis is given below.

4 Results and Discussions

The rate of occurrence of MSDs was higher in the lower back (63.1%), followed by shoulder (62.85%), knee (57.76%), neck (55.98%), hip (47.07%), upper back (44.53%), elbow and wrist (33%) and angle (25.70%) (See Fig. 2). Factors affecting MSDs in one or more body regions were age, height, weight, daily working hours, distance travelled, perceived fatigue and work experience. The outcomes of the study identify high rate of MSDs among manual working farmers and highlight the importance of work-related factors in triggering MSDs.

Pain in body parts is rated as "severe pain" followed by "somewhat hard pain", "somewhat pain", "sever discomfort" and "slight discomfort". Lower back found to experience maximum occurrence of MSDs from the obtained data, it can be concluded that the body parts experiencing higher intensity of pain among workers are hip, followed by knee, lower back, shoulder, neck, upper back, wrist, angle and elbow (See Fig. 3). From the obtained data, the usage of tools like hoe, pickaxe and shovel are the most affecting factors of MSDs, and they have the highest influence on hip compared to other body parts. Also, the usage of machete that forces bending of the worker for operation also gives strain to hip.

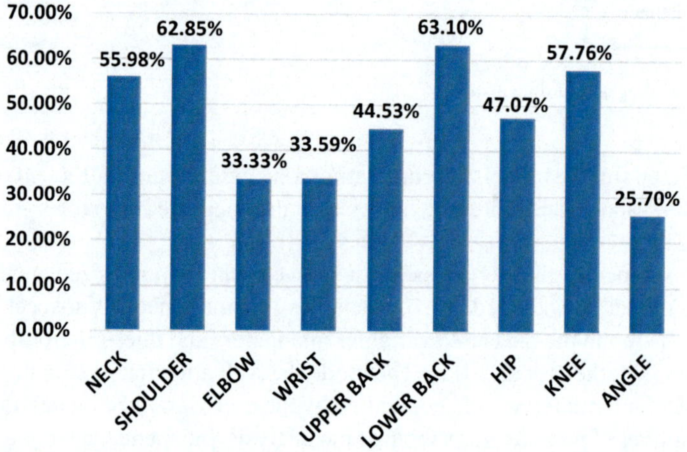

Fig. 2 Occurrence of MSDs

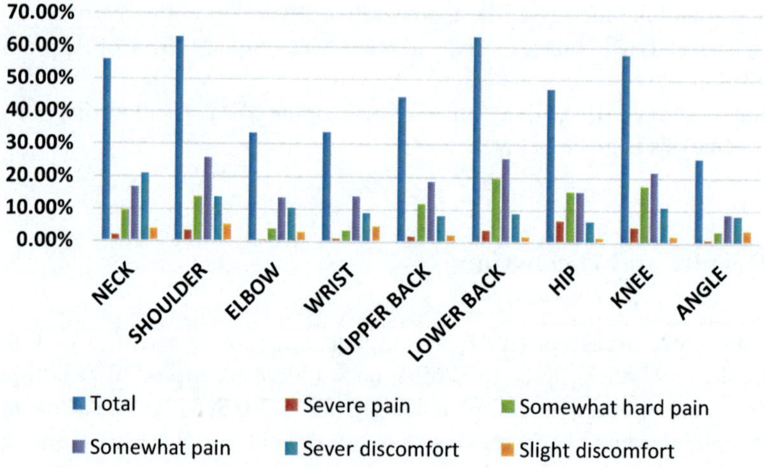

Fig. 3 Rating of MSDs

4.1 Exploratory Factor Analysis

The objective of carrying out factor analysis is to fit the variables into factors that best explain the data. It also eliminates variables that do not account for sufficient amount of the variance. The analysis is carried out on a sample of 385 data points and 33 variables after data reduction. Factor analysis is an iterative process that involves multiple attempts to get acceptable and best-fitting results.

The sampling adequacy using Kaiser-Meyer-Olkin (KMO) gives a KMO value of 0.7 or greater is considered as good while a value lower than 0.5 is considered inadequate. The result of KMO sampling adequacy test is given in Table 1. The result shows a KMO value of 0.713, which indicates that the data are adequate, and the factors extracted can explain it to an acceptable level. The Bartlett test is expected to produce significance under 0.05, and the obtained value implies that there is enough correlation between the variables to carry out a factor analysis.

Independent variables having communalities extraction greater than 0.4 are selected for the analysis. The extracted variable after PCA and factor rotation is given in Table 2.

EFA through the principal component analysis identifies 20 variables loaded into seven factors, having an Eigen value greater than one, which account for 68.202% of the total variance. An orthogonal rotation is carried out via oblimin rotation method to obtain the actual factor loadings. Here, variables having a loading under 0.3 are eliminated (Table 3).

After the factors are selected, the oblimin rotation is carried out and the pattern matrix is generated, which is given below (Table 4).

The above table shows the pattern matrix of the factors. The pattern matrix gives the factor loadings on each variable. A variable can be represented by a regression equation by taking the factor loadings as the coefficient of each factor. Factors can be termed as:

Factor 1: Contains worker characteristics like age, total experience and current experience.

Factor 2: Contains workplace characteristics like hours worked/day, days worked/week and number of hours standing/walking.

Factor 3: Contains work elements like pesticides, fertilizers and shovel.

Factor 4: Contains risk factors like working under sunlight, risk of using tool, hours of doing repetitive work.

Factor 5: Contains usage of tools like pickaxe, machete and height also falls under the same factor.

Table 1 Test statistics using KMO and Bartlett

Kaiser-Meyer-Olkin measure of sampling adequacy		0.713
Bartlett's test of sphericity	Approx. Chi-square	2558.153
	df	190
	Sig	0.000

Table 2 Communalities

Variables	Initial	Extraction
Age	1.000	0.841
Height	1.000	0.665
Total experience	1.000	0.830
Current experience	1.000	0.581
Work/day	1.000	0.773
Days/week	1.000	0.743
Stand/walk	1.000	0.570
Transport	1.000	0.831
Distance	1.000	0.818
Sunlight	1.000	0.536
Fertilizers	1.000	0.789
Pesticides	1.000	0.812
Tools	1.000	0.666
Physically exhausted	1.000	0.575
Repetitive hours	1.000	0.719
Rest	1.000	0.704
Hoe	1.000	0.595
Pickaxe	1.000	0.435
Shovel	1.000	0.590
Machete	1.000	0.567

Factor 6: Contains workplace factor like distance and transportation.
Factor 7: Contains only rest variable.

4.2 Logistic Regression

The binary logistic regression is done for the WMSD by taking the factor scores of all factors as independent variable and prevalence of WMSD as the dependent variable. WMSD carries either 1 or 0, which means that either a person has WMSD or doesn't have. This is calculated on the basis of whether the person had any severe pain in any body part every day or for more than 30 days. If the person had, then he is suffering from WMSD and given a value 1.

Table 5 gives us the coefficient of each factor in the equation. Considering the factors with significance less than 0.05, the equation can be drawn as:

$$Prevalence\,of\,WMSDs = (0.492 \times F1) + (0.351 \times F2) + (-0.471 \times F4) + (0.502 \times F5) + 1.931$$

Table 3 Total variance extracted

Component	Initial eigenvalues			Extraction sums of squared loadings			rotation sums of squared loadings
	Total	Percentage of variance (%)	Cumulative (%)	Total	Percentage of variance (%)	Cumulative (%)	Total
1	3.629	18.147	18.147	3.629	18.147	18.147	2.508
2	2.182	10.911	29.058	2.182	10.911	29.058	2.552
3	2.070	10.350	39.407	2.070	10.350	39.407	2.023
4	1.873	9.367	48.774	1.873	9.367	48.774	2.221
5	1.508	7.541	56.315	1.508	7.541	56.315	2.183
6	1.356	6.781	63.096	1.356	6.781	63.096	1.761
7	1.021	5.106	68.202	1.021	5.106	68.202	1.257
8	0.942	4.712	72.915				
9	0.821	4.107	77.022				
10	0.726	3.632	80.654				
20	0.132	0.660	100.000				

It is evident from the above equation that WMSD is mainly affected by factors $F5$, $F1$ and $F4$. $F5$ comprises of effects of tools that are Machette and Pickaxe. $F4$ comprises of Work Place Assessment variables like risk of using tool, risk of working under sunlight. Even Hoe, as a tool, has a predominant effect. The reason for the elimination of other factors is maybe due to inadequate variables covered under factor (like $F7$, which has only rest and $F6$ has only distance and transport). Moreover, $F3$ has the usage of chemical substances like pesticides and fertilizers, which is not used in all workplaces.

4.3 Structural Equation Model

The CFA is carried out using SPSS Amos software, and the hypothesized SEM is made. In this case, F1 is worker characteristic, F2 is work characteristic, F4 is risk factor and F5 is tools effect (Figs. 4 and 5).

Four factors are made, which is based on the regression equation obtained, and different indices to evaluate the fitness of the model. The results are as given below (Table 6).

As these are not falling under the limits, the model is not having a good fit. Hence, the Modification Indices are taken into consideration, and covariance between errors is given to reduce the value of Chi-square. The value of indices for this model is obtained as below (Table 7).

Table 4 Pattern matrix for the data set

	Component						
	1	2	3	4	5	6	7
Total experience	0.868						
Age	0.774						
Current experience	0.589	-0.325					
Days/week		0.823					
Stand/walk		0.737					
Work/day		0.718					
Pesticides			0.906				
Fertilisers			0.881				
Shovel	0.357		0.420				
Repetitive hours				0.782			
Hoe				−0.705			
Tools				−0.659			
Sunlight	0.317			−0.518			
Machette					−0.723		
Physically exhausted					0.666		
Height	0.377				−0.544		
Pickaxe					−0.515		
Distance						0.896	
Transport						0.882	
Rest							0.828

Table 5 Logistic regression results

Factors		B	S.E	Wald	df	Sig	Exp(B)	95% CI for EXP(B)	
								Lower	Upper
Step 1[a]	F1	0.492	0.163	9.127	1	0.003	1.636	1.189	2.251
	F2	0.351	0.148	5.596	1	0.018	1.420	1.062	1.900
	F3	−0.180	0.152	1.403	1	0.236	0.835	0.620	1.125
	F4	−0.471	0.150	9.933	1	0.002	0.624	0.466	0.837
	F5	0.502	0.173	8.471	1	0.004	1.653	1.178	2.318
	F6	0.138	0.156	0.783	1	0.376	1.148	0.845	1.560
	F7	−0.174	0.176	0.977	1	0.323	0.841	0.596	1.186
	Constant	1.931	0.172	126.548	1	0.000	6.894		

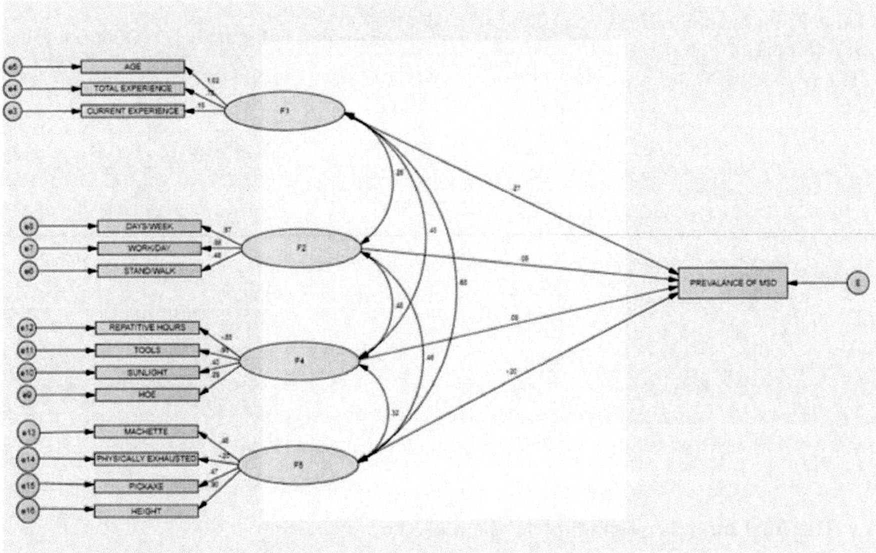

Fig. 4 Initial model

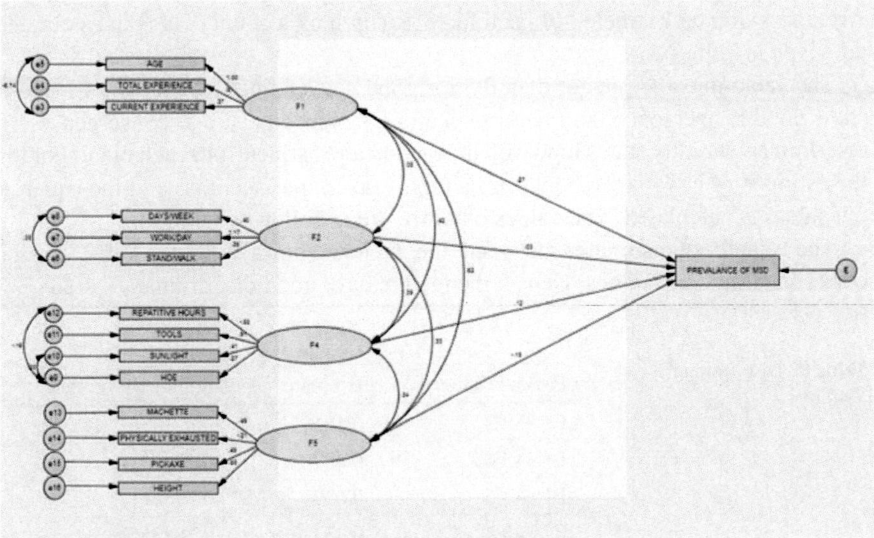

Fig. 5 Final model

Table 6 Fit Indices for final SEM model

Fit indices	Obtained value	Fit indices	Obtained value
Chi sq	487.635	NFI	0.731
GFI	0.852	TLI	0.692
AGFI	0.780	RMSEA	0.114
CFI	0.762		

Table 7 Fit Indices for Final SEM Model

Fit indices	Obtained value	Fit indices	Obtained value
Chi sq	372.109	NFI	0.795
GFI	0.891	TLI	0.765
AGFI	0.830	RMSEA	0.100
CFI	0.828		

Table 8 Convergent validity

Factors	AVE
F1	0.55925
F2	0.620667
F4	0.449995
F5	0.421874

The final model obtained is as given below:

These values are quite closer to 1 and RMSEA is also reduced, hence this value gives a comparatively better fit. For calculating the validity of the model, convergent and discriminant validity are calculated. Convergent validity is calculated by taking Average Variance Extracted of each factor taken into account. The values obtained are given in Table 8.

The value must be greater than 0.5 for good convergence, the factors F1 and F2 show good convergence, and factors F3 and F4 show an average convergence. The discriminant validity states how well the factors are distinct from each other. For this, the estimate of covariance between factors is taken into account, and the square of the values is calculated. The values obtained are as follows (Table 9).

The square of estimates between the factors shows a lower value than its corresponding AVE values. Hence, the factors have good discriminant validity.

Table 9 Discriminant Validity

Factor relation	Estimate	Square of estimate
F1 ↔ F2	0.166	0.027556
F1 ↔ F4	0.429	0.184041
F1 ↔ F5	0.629	0.395641
F2 ↔ F4	0.434	0.188356
F2 ↔ F5	0.373	0.139129
F4 ↔ F5	0.32	0.1024

5 Conclusions and Scope for Future Works

The study shows the evidence for the prevalence of musculoskeletal disorders among women agriculture workers of Kerala. The reason for this issue is that the farm-workers are exposed to various physical risks, and the existing tools employed are primarily intended for men agriculture workers. Out of the 436 women studied, 416 reported to have some sort of musculoskeletal discomfort. The body parts of workers who encountered discomfort are lower back (63.10%) region, followed by shoulder (62.85%) and knee (57.76%), and 45% of workers have encountered with injury during the work. From the survey report, most of the workers are experiencing some sort of trouble (ache, pain, numbness, discomfort, etc.) for the past period of 12 months.

The regression model obtained is interpreted and found out that the usage of tool is the most predominant reason for the cause of WMSD. The usage of machete and pickaxe is the main reason for the cause. Another factor like long hours of working under sunlight is also a cause for WMSD; this is because workers become more fatigue after working for 8 hrs on an average and become physically exhausted. The usage of pickaxe and height fall under the same factor indicating that the pickaxe usage is causing WMSD for tall workers. From the study, some recommended measures suggested are listed below:

- Rests/pauses during the full day working encourage workers to work efficiently.
- Muscle relaxing exercises should be included within the resting break to prevent pain.
- For reducing the problem faced by farmers in various postures, the possibility of providing efficient and ergonomically designed tools/equipment is to be considered.
- For updating the knowledge about safety and health among agricultural workers, appropriate educational/training camps can be employed.

The study utilizes data collected from six districts only, the data from entire districts can provide more accurate results regarding WMSDs among agricultural population and the study is not verifying whether there is a significant difference in disorders existing in the agricultural workers of different locations in Kerala.

The study opens the door for some future works. The study can be extended to the cases like:

- The musculoskeletal caused by the specific tool used by women workers.
- Selection of suitable tools.

This study has been conducted in State of Kerala, having the highest literacy rate in India and the lowest number of women agricultural workers compared to other states. Women workers from other states can be included in future studies so that more advancement can be implemented.

References

Adewusi S, Rakheja S, Marcotte P, Thomas M (2013) Distributed vibration power absorption of the human hand-arm system in different postures coupled with vibrating handle and power tools. Int J Ind Ergon 43(4). https://doi.org/10.1016/j.ergon.2013.04.004

Agbor AM, Hilbert K (2014) Work-related musculoskeletal disorders amongst oral health workers in cameroon. J Oral Health Dental Manag 15(6):1–6

Albers J, Estill C, MacDonald L (2005) Identification of ergonomics interventions used to reduce musculoskeletal loading for building installation tasks. Appl Ergon 36(4):427–439. https://doi.org/10.1016/j.apergo.2004.07.005

Azmir NA, Ghazali MI, Yahya MN, Ali MH, Song JI (2015) Effect of hand-arm vibration on the development of vibration induce disorder among grass cutter workers. Proced Manuf 2:87–91. https://doi.org/10.1016/j.promfg.2015.07.015

Bahri S (2014) Occuptional safety and health in commodity agriculture . Case studies from Malaysian agricultural perspective. CRC Press

Basahel AM (2015) Investigation of work-related musculoskeletal disorders (MSDs) in warehouse workers in Saudi Arabia. In: 6th International conference on applied human factors and ergonomics (AHFE 2015) and the affiliated conferences, vol. 3, pp 4643–4649. https://doi.org/10.1016/j.promfg.2015.07.551

Basahel AM (2016) Physical demand, fatigue and shift work in the installation and maintenance of window air-conditioner units. Life Sci J 13(5):46–55. https://doi.org/10.7537/marslsj13051605. Keywords

Bengal W, Gangopadhyay S, Das B, Das T, Ghoshal G (2005) An ergonomic study on posture-related discomfort among preadolescent agricultural workers of West Bengal, India. Int J Occup Saf Ergnom 11(3):315–322. https://doi.org/10.1080/10803548.2005.11076652

Borah S (2015) Physiological workload of hill farm women of Meghalaya , India involved in firewood collection. Proced Manuf 3:4984–4990. Elsevier B.V. https://doi.org/10.1016/j.promfg.2015.07.648

Burdorf A, Sorock G (1997) Positive and negetive evidence of risk factors for back disorders. Scand J Work Environ Health 23(4):243–256

Burdorf A, Govaert G, Elders L (1991) Postural load and back pain of workers in the manufacturing of prefabricated concrete elements. Ergonomics 34(7):909–918. https://doi.org/10.1080/00140139108964834

Cherian T, Rakheja S, Bhat RB (1996) An analytical investigation of an energy flow divider to attenuate hand-transmitted vibration. Int J Ind Ergon 17(6):455–467

Choobineh A, Rajaeefard A, Neghab M (2006) Association between perceived demands and musculoskeletal disorders among hospital nurses of Shiraz University of Medical Sciences: a questionnaire survey association between perceived demands and musculoskeletal disorders among hospital nurses of Shiraz. Int J Occup Saf Ergon 12(4):409–416. https://doi.org/10.1080/10803548.2006.11076699

Ciriello VM, Snook SH (1999) Survey of manual handling tasks. Int J Ind Ergon 23(3):149–156

Das B, Ghosh T, Gangopadhyay S (2012) Assessment of ergonomic and occupational health-related problems among female prawn seed collectors of Sunderbans, West Bengal, India. Int J Occup Saf Ergon JOSE 18(4):531–540. https://doi.org/10.1080/10803548.2012.11076949

Das B, Ghosh T, Gangopadhyay S (2013) Prevalence of musculoskeletal disorders and occupational health problems among groundnut farmers of West Bengal, India. J Hum Ergol 42(1–2):1–12. https://doi.org/10.1539/joh.12-0185-OA

Das B, Gangopadhyay S (2011) An ergonomics evaluation of posture related discomfort and occupational health problems among rice farmers 10:25–38. https://doi.org/10.3233/OER-2011-0190

Dempsey PG (2003) A survey of lifting and lowering tasks. Int J Ind Ergon 31:11–16

Devereux JJ, Vlachonikolis IG, Buckle PW (2002) Epidemiological study to investigate potential interaction between physical and psychosocial factors at work that may increase the risk of symptoms of musculoskeletal disorder of the neck and upper limb. Occup Environ Med 59(4):269–277. http://search.ebscohost.com/login.aspx?direct=true&db=cin20&AN=106974859&site=ehost-live

Dewangan KN, Tewari VK (2010) Handle grips for reducing hand-transmitted vibration in hand tractor. Int Agric Eng J 19(2):48–57

Dhimmar V, Sheth S, Shah S (2011) Proposed design of workstation for rural blacksmith. Natl Conf Recent Trend Eng Technol 1–4

Fathallah FA (2010) Musculoskeletal disorders in labor-intensive agriculture. Appl Ergon 41(6):738–743. https://doi.org/10.1016/j.apergo.2010.03.003

Field A (2013) Discovering statistics using IBM SPSS (4th edn). SAGE Publications Ltd.

Gangopadhyay S, Das B, Das T, Ghoshal G (2015) An ergonomics study on posture-related discomfort and occupational-related disorders among stonecutters of West Bengal, India. Int J Occup Saf Ergon 3548(1):69–79. https://doi.org/10.1080/10803548.2010.11076830

Gupta G (2013) Prevalence of musculoskeletal disorders in farmers of Kanpur-Rural, India. J Community Med Health Educ 3(7):249–253. https://doi.org/10.4172/2161-0711.1000249

Hair JF, Black WC, Babin BJ, Anderson RE (2010) Multivariate data analysis. Vectors (13th edn). https://doi.org/10.1016/j.ijpharm.2011.02.019

Halldin M, Rosell M, Faire UD (2007) The metabolic syndrome: prevalence and association to leisure-time and work-related physical activity in 60-year-old men and women. Nutr Metab Cardiovasc Dis 17(5):349–357. https://doi.org/10.1016/j.numecd.2006.01.002

Hamilton LC (2013) Stata structural equation modeling reference manual (13th edn). A Stata Press Publication

Hoefkens C, Pieniak Z, Van Camp J, Verbeke W (2012) Explaining the effects of a point-of-purchase nutrition-information intervention in university canteens: a structural equation modelling analysis. Int J Behav Nutr Phys Act 9:1–10. https://doi.org/10.1186/1479-5868-9-111

International Labour Organization (2001) Agriculture: a hazardous work. http://www.ilo.org/safework/areasofwork/hazardous-work/WCMS_110188/lang--en/index.htm. Accessed 6 June 2020

Jager M, Luttmann A (1999) Critical survey on the biomechanical criterion in the NIOSH method for the design and evaluation of manual lifting tasks. Int J Ind Ergon 23(4):331–337

Kolstrup CL (2012) Work-related musculoskeletal discomfort of dairy farmers and employed workers. J Occup Med Toxicol 7(1):23–32. https://doi.org/10.1186/1745-6673-7-23

Lee Y-H, Chen Y-L (1996) Strength-limiting factors for the determination of frequent and infrequent lifting capacities. Int J Ind Ergon 18(1):51–60. https://doi.org/10.1016/0169-8141(95)00028-3

Lim C, Kong Y (2014) Effects of the resting time associated with the number of trials on the total and individual finger forces in a maximum grasping task. Appl Ergon 45(3):443–449. https://doi.org/10.1016/j.apergo.2013.06.002

Litchfield MH (1999) Agricultural work-related injury and Ill-health and the economic cost. Environ Sci Pollut Res Int 6(3):175–182. https://doi.org/10.1007/BF02987623

Majumder J (2014) Healthy workplace intervention: scope of basic occupational health services (BOHS) in informal occupations in India. Asian-Pacific News Lett Occup Health Saf 21(8):53–55

Mandal BB, Srivastava AK (2006) Risk from vibration in Indian mines. Indian J Occup Environ Med 10(2):53–57

Mehta CRÃ, Gite LP, Pharade SC, Majumder J, Pandey MM (2008) Review of anthropometric considerations for tractor seat design. Int J Ind Ergon 38(5):546–554. https://doi.org/10.1016/j.ergon.2007.08.019

Meyers JM, Miles JA, Tejeda DG, Faucett J, MFS J, Weber E, … Garcia L (2008) Priority Risk Factors for Back Injury in agricultural field work. J Agromed 8(1):39–54. https://doi.org/10.1300/J096v08n01_05

Mobed K, Gold EB, Schenker MB (1987) Occupational health problems among migrant and seasonal farm workers. West J Med 157(3):367–374

Moon SD, Sauter SL (2005) Beyond biomechanics: psychosocial aspects of musculoskeletal disorders in office work. CRC Press, Taylor and Francis Group

Nimbarte AD (2014) Risk of neck musculoskeletal disorders among males and females in lifting exertions. Int J Ind Ergon 44(2). https://doi.org/10.1016/j.ergon.2013.01.007

Niu S (2000) Ergonomics and occupational safety and health: an ILO perspective. Appl Ergon 41(6):744–753

Osborne A, Blake C, McNamara J, Meredith D, Phelan J, Cunningham C (2010) Musculoskeletal disorders among irish farmers. Occup Med 60(8):598–603. https://doi.org/10.1093/occmed/kqq146

Osborne A, Blake ÃC, Fullen BM, Meredith D, Phelan J, Mcnamara J, Cunningham C (2012a) Risk factors for musculoskeletal disorders among farm owners and farm workers: a systematic review. Am J Ind Med Ind Med 55(4):376–389. https://doi.org/10.1002/ajim.22001

Osborne A, Blake ÃC, Fullen BM, Meredith D, Phelan J, Mcnamara J, Cunningham C (2012b) Risk factors for musculoskeletal disorders among farm owners and farm workers . Syst Rev 389:376–389. https://doi.org/10.1002/ajim.22001.

Palmer KT, Syddall H, Cooper C, Coggon D (2003) Smoking and musculoskeletal disorders: findings from a British National survey. Ann Rheum Dis 62(1):33–36. https://doi.org/10.1136/ard.62.1.33

Parikh AG, Schulze LJH, Chen JG (Jacob), Cleveland T (1997) Maximum acceptable weight of asymmetrical lifting and lowering of postal sacks. Int J Ind Ergon 19(1):1–7. https://doi.org/10.1016/0169-8141(94)00113-8

Parkes KR, Carnell S, Farmer E (2005) Musculo-skeletal disorders, mental health and the work environment (research report 316/2005). Health and safety executive books, Sudbury, UK

Rahmatalla S, Deshaw J (2011) Effective seat-to-head transmissibility in whole-body vibration: effects of posture and arm position. J Sound Vib 330(25):6277–6286. https://doi.org/10.1016/j.jsv.2011.07.034

Ranasinghe P, Perera YS, Lamabadusuriya DA, Kulatunga S, Jayawardana N, Rajapakse S, Katulanda P (2011) Work related complaints of neck, shoulder and arm among computer office workers: a cross-sectional evaluation of prevalence and risk factors in a developing country. Environ Health 10(1):70. https://doi.org/10.1186/1476-069X-10-70

Rosecrance J, Rodgers G, Merlino L (2006) Low back pain and musculoskeletal symptoms among kansas farmers. Am J Ind Med 556(7):547–556. https://doi.org/10.1002/ajim.20324

Salas EA, Vi P, Reider VL, Moore AE (2016) Factors affecting the risk of developing lower back musculoskeletal disorders (MSDs) in experienced and inexperienced rodworkers. Appl Ergon 52:62–68. https://doi.org/10.1016/j.apergo.2015.06.016

Shojaei I, Vazirian M, Croft E, Nussbaum MA, Bazrgari B (2016) Age related differences in mechanical demands imposed on the lower back by manual material handling tasks. J Biomech 49(6):896–903. https://doi.org/10.1016/j.jbiomech.2015.10.037

Singh S, Mathur P, Rathore M (2007) Weeders for drudgery reduction of women farm workers in India. J Agric Eng 44(3):33–38

Song J, Qu X (2014) Age-related biomechanical differences during asymmetric lifting. Int J Ind Ergon 44(5). https://doi.org/10.1016/j.ergon.2014.06.008

Srinavin K, Mohamed S (2003) Thermal environment and construction workers ' productivity: some evidence from Thailand. Build Environ 38:339–345

Srinivas T, Ramanathan S (2005) A study on economic analysis of elephant foot yam production in India. Agric Econ Res Rev 18:241–252

Stanbury M, Rafferty AP, Rosenman K (2008) Prevalence of hearing loss and work-related noise-induced hearing loss in michigan. J Occup Environ Med 51(1):72–79. https://doi.org/10.1097/JOM.0b013e31815b568c

Tai SHC, Tam JLM (1997) A lifestyle analysis of female consumers in greater China. Psychol Mark 14(3):287–307. https://doi.org/10.1002/(SICI)1520-6793(199705)14:3%3c287::AID-MAR5%3e3.0.CO;2-7

Tan C, Delbressine F, Chen W, Rauterberg M (2008) Subjective and objective measurements for comfortable truck driver's seat. In: Proceedings of 9th international symposium on advanced vehicle control (AVEC 2008), pp 851–856

Tanaka S, Petersen M, Cameron L (2001) Prevalence and risk factors of tendinitis and related disorders of the distal upper extremity among U S workers: comparison to carpal tunnel syndrome. Am J Ind Med Ind Med 39(3):328–335

The World Bank (n.d.) Employment in agriculture. https://data.worldbank.org/indicator/SL.AGR.EMPL.FE.ZS. Accessed 23 March 2018

Van Der Beek AJ, Erik S, Windhorst J, Burdorf A (2005) An eevaluation of methods assessing the physical demands of manual lifting in scaffolding. Appl Ergon 36(2):213–222. https://doi.org/10.1016/j.apergo.2004.10.012

Visser S, van der Molen HF, Kuijer PPFM, Hoozemans MJM, Frings-Dresen MHW (2014) Evaluation of team lifting on work demands, workload and workers' evaluation: an observational field study. Appl Ergon 45(6). https://doi.org/10.1016/j.apergo.2014.05.009

Waghorn G, Chant D, Lloyd C (2006) Labor force activity among australians with musculoskeletal disorders comorbid with depression and anxiety disorders. J Occup Rehabil 16(2):241–252. https://doi.org/10.1007/s10926-006-9018-3

Walker-Bone K, Palmer KT (2002) Musculoskeletal disorders in farmers and farm workers. Occup Med 52(8):441–450. https://doi.org/10.1093/occmed/52.8.441

Wang W, Rakheja S, Boileau PÉ (2004) Effects of sitting postures on biodynamic response of seated occupants under vertical vibration. Int J Ind Ergon 34(4):289–306. https://doi.org/10.1016/j.ergon.2004.04.009

World Health Organization (2004) Occupational and work-related diseases. https://www.who.int/occupational_health/activities/occupational_work_diseases/en/. Accessed 1 July 2020

Xiang H, Stallones L, Keefe TJ (1999) Back pain and agricultural work among farmers: an analysis of the colorado farm family health and hazard surveillance survey. Am J Ind Med 35(3):310–316. https://doi.org/10.1002/(SICI)1097-0274(199903)35:3%3c310::AID-AJIM12%3e3.0.CO;2-D

Xu XS, Dong RG, Welcome DE, Warren C, McDowell TW, Wu JZ (2016) Vibrations transmitted from human hands to upper arm, shoulder, back, neck, and head. Int J Ind Ergon 30(1):1–12. https://doi.org/10.1016/j.ergon.2016.07.001

Yarnell J, Yu S, McCrum E, Arveiler D, Haas B, Dallongeville J, … Cambien F (2005) Education, socioeconomic and lifestyle factors, and risk of coronary heart disease: the PRIME study. Int J Epidemiol 34(2):268–275. https://doi.org/10.1093/ije/dyh267

Chapter 7
Gender-Specific Agricultural Tool for Enhanced Productivity and Human Health

P. Parimalam, P. K. Padmanathan, S. Logeswari, and B. Nallakurumban

1 Introduction

Agriculture is the major occupation that involves various operations and crop-specific production practices in different regions. As per FAO (2011) report, about 50% of World Food is produced by women. In addition to farming activities, women also shoulder the responsibility of household activities and child care. In agriculture, women farmers are more involved in activities that are strenuous and time-consuming. Women contribute to about three-fourths of the labour required for agricultural operations (Mishra 2013). The drudgery of farm women is not addressed specifically by society. Tools developed to till date in agriculture address the issues of farm mechanization focusing on activities performed by men. Higher participation of the female workforce and the changing scenario of farm technologies demand more emphasis on a gender-friendly tool (Mehta et al. 2018). The major activities which are timely to be performed in farming such as transplanting, weeding and harvesting are exclusively done by women farmers (Aggarwal et al. 2013). Presently, most of the operations under vegetable cultivation are accomplished manually. These operations are accomplished in kneeling posture or squatting posture (Meyers et al. 2015). Therefore, they are more tedious, uncomfortable, tiresome and drudgery prone. Transplanting is the most effective cultivation process that would benefit from farm mechanization in vegetable production. Vegetable transplantation in India is mainly accomplished manually and requires 260–320 man-hours/ha, which is labour-intensive, expensive, time-consuming and often results in non-uniform plant distribution (Dixit et al. 2018; Kumar and Raheman 2008; Manes

P. Parimalam (✉) · S. Logeswari · B. Nallakurumban
Department of Family Resource Management and Consumer Science, Community Science College & Research Institute, TNAU, Madurai, Tamil Nadu, India

P. K. Padmanathan
Department of Farm Machinery, Agricultural Engineering College, TNAU, Kumulur, Tamil Nadu, India

© The Author(s), under exclusive license to Springer Nature Singapore Pte Ltd. 2022, corrected publication 2022

D. Bijulal et al. (eds.), *Technology-Enabled Work-System Design*, Design Science and Innovation, https://doi.org/10.1007/978-981-16-1884-0_7

et al. 2010). Outstanding developments in agriculture operations came into existence after the mechanization of agricultural operations, and this led to increased output and reduced the level of burden at work (Singh 2018). Development and research activities in India have made some attempts in recent years to develop transplanters sustainable for our conditions. Based on the ergonomic and mechanical considerations there is a necessity to incorporate refinements in the tools used by farm women, to improve the fit between the physical demand of the tool and the worker (Durga 2018). The transplantation of seedlings is a time-consuming and labour-intensive task.

2 Methodology

2.1 *Survey of Farm Women*

An exploratory research survey through interview method was adopted to conduct the study in four districts of Tamil Nadu, namely, Madurai, Theni, Viruthunagar and Trichy. A total of 368 farm women were randomly selected for the study, and information regarding activities performed by them and the need for gender-friendly tools were recorded in the study. Based on the survey results, activities performed by farm women have been prioritized to address their field-level problems.

2.2 *Prototype Fabrication of Handheld Seedling Transplanter*

A prototype handheld seedling transplanter was designed based on the anthropometric dimension of the user population and also on the seedling parameters. The seedlings parameters such as the floral spread (canopy) of the seedlings and the height of the seedlings were the major parameters of the seedlings to design the mouth of the jaw and the diameter of the seedling tube. The anthropometric dimensions of the farm women and design consideration as cited by Gite et al. (2009) were referred for the designing of the seedling transplanter (Fig. 1) presents the transplanter used for the study

The specification of the handheld seedling transplanter is given in Table 1.

2.3 *Evaluation of Handheld Seedling Transplanter*

Twelve farm women with work experience of more than 10 years in farming activities were purposively selected in the study on evaluation of the tool for transplanting activity. All the subjects performed both methods of transplanting, i.e. conventional

Fig. 1 Handheld seedling transplanter

Table 1 Specifications of the handheld seedling transplanter

S. no	Features	Specifications
1	Weight of the tool (kg)	1.65
2	Height of the transplanter (mm)	900
3	Height of handle from the ground (mm)	820
4	Length of the handle on the left side (mm)	115
5	Length of the handle on the right side (mm)	160
6	Diameter of the handle (mm)	20
7	Diameter of the seedling hopper (mm)	65
9	Length of the jaw (mm)	130
10	Apex angle of the jaw	37.5
11	Length of the lever (mm)	690
12	Spacing marker (mm)	300–450

and by using handheld seedling transplanter. The experiments were carried out for all the selected subjects. Work experience and free from health ailments of the subjects were the main inclusion criteria for selecting the subjects. Oral consent from the subjects was taken prior to the conduct of the experiment.

The efficiency of the tool was tested in 0.1 ha area for each method, i.e. conventional and by using handheld seedlings transplanter. The seedlings of 18–25 days old raised portray were used. The average seedling height was 150 mm. Portray seedlings were irrigated one day before the transplanting. Seedlings of the vegetable and flower crops were used for transplanting using the designed tool.

Measurement of heart rate and energy expenditure

A portable Polar heart rate monitor (V800) was used to measure the heart rate during the transplanting operation. Before starting an experiment all functions of the monitor were set as per the instructions given in the manual. Sufficient rest was given to the subject before starting the experiment to make the heart rate stable. The experiment was carried out after recording the resting heart rate (RHR). While conducting the experiment the working heart rate of the subjects was recorded continuously. On completion of the transplanting activity, the recovery heart rate was recorded. From the downloaded data, the value of heart rate was taken as the physiological responses of the subjects. The energy expenditure of the subjects was derived using the Varghese et al. (1994) equation from the heart rate method.

Data analysis

Data analysis and interpretation were carried out using the SPSS-16.0 version. Percentage and paired t-test have been used to arrive at valid conclusions.

3 Results and Discussion

The respondents involved in the study were all farm women residing in Tamil Nadu. The sociological profile of the farm women such as age, education qualification, family type and work experience are discussed in Table 2.

The mean age of farm women in the study was 43 $\pm$ 10.97 years. Among them, 41% of the farm women belong to the late middle age group followed by 30% of young age farm women and 25% middle age farm women. The present results were in concurrence with the results of Srivastava and Singh (2014) as they had reported that approximately half of the women involved in agriculture activities belong to the middle-aged group. Regarding the educational level of the farm women, 63% were literate. Among the literate farm women, 43% of them had completed primary education, followed by 16% of them had completed their high school education. Only 4% were graduates; 65% of them were belonging to the nuclear family type. The work experience of farm women revealed the expertise of farm women in agriculture. The mean work experience of farm women was 19 $\pm$ 9.7 years. 68% of the farm women had experience in agriculture occupation for more than 10 years. Among them, 39% of the women were in agricultural operations for 10–20 years and 21% of them had been in agriculture for 20–30 years. Only 8% of the farm women had more than 30 years of experience in agriculture (Table 3).

Agricultural operations range from land preparation to harvesting and post-harvest operations, in which women are involved. Figure 2 presents the participation of women in different agricultural activities. Weeding, harvesting and transplanting

Table 2 Sociological profile of the farm women

S. No	Particulars	No. of respondents (n = 368)	Percentage (%)
1	Age		
	Young age (<35 years)	109	30
	Middle age (35–45 years)	94	25
	Late middle-aged (45–59 years)	148	41
	Old age (60 and above)	17	4
2	Education		
	Illiterate	137	37
	Primary education	157	43
	High school	58	16
	Graduation	16	4
3	Family type		
	Nuclear	241	65
	Joint	127	35
4	Work experience (years)		
	Less than 10	118	32
	10–20	144	39
	20–30	78	21
	30–40	22	6
	More than 40	6	2

Table 3 Problems reported by farm women during transplanting

S. no	Problems	No. of farm women (n = 368)	Percentage (%)
1	Lack of tool	122	33
2	Bent posture results in low back pain	109	30
3	Time-consuming activity	95	26
4	Muscle cramps in leg and knee	56	15

were predominantly women-based activities. A similar trend of participation of women has been reported in studies done by Chaudhuri et al. (2002), Singh (2008), Parish (2005). Sixty six per cent of them performed transplanting as the major activity, and a similar trend was reported in earlier studies by Nain and Kumar (2010).

Farm mechanization is adopted when the cultivable area is large, while in small and fragmented landholdings, farmers prefer to do most of the agricultural activities using labourers. Lack of knowledge/awareness about the availability of tools was stated by 33% of the farm women. Musculoskeletal discomforts in different body parts have been expressed by one-third of the farm women. Women spend long hours with

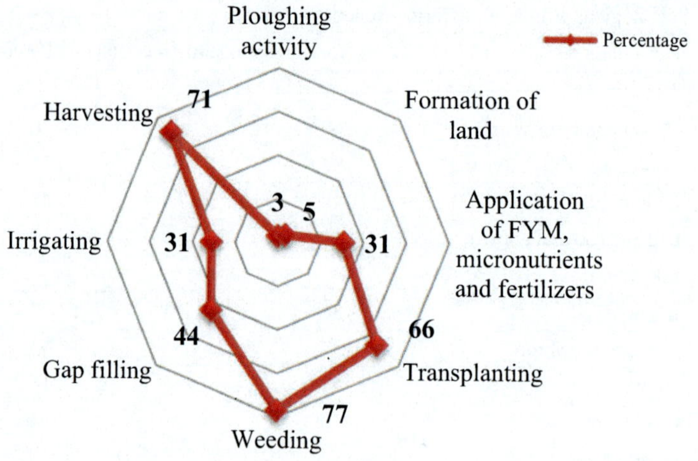

Fig. 2 Women participation in agricultural operations

much labour in respective operations resulting in fatigue and drudgery (Parimalam and Meenakshi 2017). Manual transplanting is a time-consuming, laborious task that needs alteration to improve human efficiency and productivity. Table 3 presents the problems faced by women in transplanting activities.

Transplanting operations which are performed manually account for 40% of total working hours of cultivation. The traditional practice is to hold a bunch of seedlings/seedling tray in one hand and separate seedlings in the other hand and press down the roots in the soil with bare hands. Several reports state that 300–320 man-hours are required for transplanting one hectare of landholding (Figs. 3 and 4).

Transplanting using the handheld seedling transplanter tool facilitates farm women to transplant seedlings in a standing posture. The seedling has to be dropped inside the seedling tube after pressing the seedling tube inside the soil. Then the handle lever is operated so that the seedlings inside the seedling pipe get pushed inside the soil and while releasing the handle lever the mouth of the jaw gets closed, thereby closing the bottom area of the seedling by sand after transplanting. This operation involves farm women to transplant seedlings in a standing posture and aids in the reduction of forward bending which leads to back pain during conventional transplanting. The use of the handheld seedling transplanter avoids postural discomfort and helps to cover a large area for planting seedlings (Times of India, Sep 2018).

The physical characteristics of the selected subjects are given in Table 4. All the 12 subjects had considerable experience in agricultural activities and a majority of the participants had normal BMI.

The heart rate and energy expenditure of the farm women while using the handheld seedling transplanter against the conventional method are discussed in Table 5.

Heart rate is the major parameter that indicates the physiological workload that the subject experienced while performing activities. The mean resting heart rate

Fig. 3 Conventional method of transplanting seedlings

Fig. 4 Farm women using handheld seedling transplanter

Table 4 Details of selected farm women

S. no	Particulars	Mean	S.D
1	Age (years)	41	4.75
2	Weight (kg)	52.7	7.01
3	Height (cm)	156.8	3.49
4	BMI	21.4	2.49
5	Work experience (years)	17	3.13

Table 5 Heart rate and energy expenditure of farm women while transplanting seedlings

S. no	Particulars	Conventional method	Handheld seedling transplanter	Paired t-test
1	Resting heart rate (bpm)	85 ± 5.37	84 ± 5.85	–
2	Working heart rate (bpm)	117 ± 9.89	107 ± 11	4.577**
3	Energy expenditure (kJ/min)	10 ± 1.58	8 ± 1.49	6.938**

** $P < 0.01$ at 99% confidence interval

of the 12 subjects was 85 bpm and while performing the conventional method of transplanting in a bent posture and the working heart rate was 117 bpm. Similarly, the heart rate of subjects while performing transplanting activity using the handheld seedling transplanter was 107 bpm. According to Dixit et al. (2018), the working heart rate while transplanting using a transplanter was 112 bpm. The energy expenditure was also analysed for the conventional method and for working with the handheld seedling transplanter. This showed that the energy expenditure was 10 kJ/min for conventional and 8 kJ/min while using the tool which is similar to the study conducted by Gupta and Sharma (2017) on evaluation of physiological workload of farm women while transplanting vegetable seedlings. There was a highly significant reduction ($P < 0.01$) in the working heart rate and energy expenditure of the subjects while using the handheld seedling transplanter (Fig. 5).

The productivity of the subjects was analysed during the conventional method and also by using handheld seedling transplanter for transplanting. The number of seedlings transplanted per minute while transplanting seedlings by the conventional method was 17 and while using the tool it was 21. A manually operated single row vegetable transplanter developed by Nandede et al. (2017) reported 16 seedlings/minute can be transplanted. There was a significant increase in the number of transplanted seedlings while using the handheld seedling transplanter.

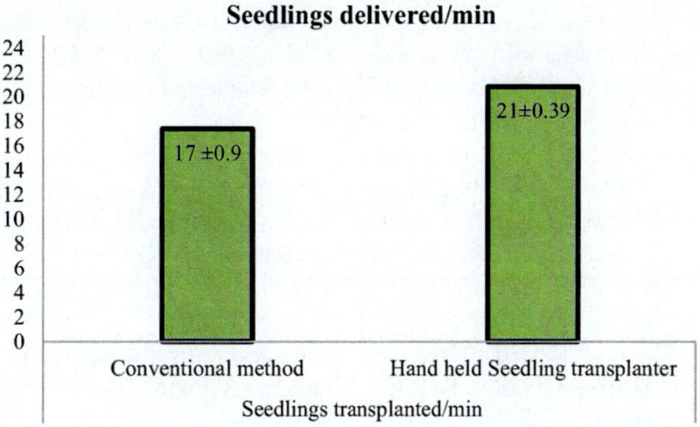

Fig. 5 Number of seedlings delivered per minute

4 Conclusion

Indian agriculture involves women in major operations but still lacks in gender-specific tools that are region-specific according to the soil and crop. The study conducted in the southern regions of Tamil Nadu revealed that middle-aged women are the major agricultural workers. The three major activities involving women were transplanting, weeding and harvesting. Problems faced by women in transplanting seedlings paved the way for the development of handheld seedling transplanters. The transplanter was designed based on the physical characteristics of the seedling and the anthropometric dimensions of the women. Consented efforts are to be made to develop tools suitable for farm women and also to make them accessible to enhance their productivity and improve their health. Hence, the development of a tool that is gender-specific and region-specific will lead to the enhanced productivity of women workers in agriculture.

References

Aggarwal H, Sharma S, Sharma R (2013) A study of agricultural activities performed by rural women and problems faced by them in Jammu district of J & K state. Int J Sci Res Publ 3(1):1–3

Chaudhuri D, Singh VV, Dubey AK (2002) Refinement and adoption of mechanical vegetable transplanter for Indian condition. Agric Eng Today 26(5–6):11–20

Dixit J, Kumar V, Ali M (2018) Development and evaluation of a single row manual vegetable transplanter. Agric Eng Today 58–66

Durga MPL (2018) Dissertation on design and development of mini tractor drawn transplanter for vegetable crops. ANGRAU 155–168

FAO (2011) The state of food and agriculture. Rome

Gite LP, Majumder J, Mehta CR, Khadatkar A (2009) Anthropometric and strength data of Indian agricultural workers for farm equipments design. AICRP on Ergonomics and safety in agriculture. CIAE, Bhopal

Gupta R, Sharma A (2017) Sapling transplanter: an effective tool for drudgery mitigation of women in vegetable production system. Asian J Home Sci 12(2):609–613

Khadatkar A, Mathur SM, Gaikwad BB (2018) Automation in transplanting: a smart way to vegetable cultivation. Current Sci 115(10). 1884–1892

Kumar GVP, Raheman H (2008) Vegetable transplanters for use in developing countries: a review. Int J Veg Sci 14(3):232–255

Manes GS, Dixit AK, Sharda A, Singh S, Singh K (2010) Development and evaluation of tractor operated vegetable transplanter. Agric Mech Asia Africa Latin Am 41(3):89–92

Mehta CR, Gite LP, Khadatkar A (2018) Women empowerment through agricultural mechanization in India. Curr Sci 114(9):1934–1940

Meyers JM, Fathallah FA, Janowitz I (2015) Stooped and squatting postures in the workplace. Natl Ag Saf Database 2–7

Mishra P (2013) A study on participation of rural women in agriculture activities in Morar block District Gwalior. Rajmata Vijayaraje Scindia Krishi Vishwa Vidyalaya College of Agriculture. Gwalior, 22–28

Nain MS, Kumar P (2010) A study of women participation and decision making in farm management. J Community Mobiliz Sustain Dev 5(1):67–71

Nandede BM, Carpenter G, Byale NA, Rudragouda C, Jadhav ML, Pagare V (2017) Manually operated single row vegetable transplanter for vegetable seedlings. 9(53):4911–4913

Parimalam P, Meenakshi J (2017) An ergonomic evaluation of grain picking activity of farm women. Adv Res J Soc Sci 8(1):1–8

Parish RL (2005) Current developments in seeders and planters for vegetable crops. Horticu Technol 15(2):1–6

Singh S (2008) Research highlights. All India coordinated research project on farm implements and machinery, central institute of agricultural engineering (CIAE). Bhopal, Madhya Pradesh, India

Singh AK (2018) Dissertation on performance evaluation of two row semi automatic vegetable transplanter. BAU 144–159

Srivastava S, Singh B (2014) Understanding nutritional situation of farm women in rural arid areas of Rajasthan: a case study. J Agric Life Sci 1(2):17–20

Times of India (2018) Article on TNAU designs tool to help female farm workers. Published on 26th September 2018. http://timesofindia.indiatimes.com/articleshow/65956845.cms?utm_source=contentofinterest&utm_medium=text&utm_campaign=cppst

Varghese MA, Saha PN, Atreya N (1994) A rapid appraisal of occupational workload from a modified scale of perceived exertion. Ergonomics 37:487–491

Chapter 8
Digital Human Modelling and Ergonomic Assessment of Handedness in Street Sweeping

A. Ajay Vishnu, K. Shashank Siddaveeraiah, K. Venkata Rahul, S. Nirmal, Priyanka Thalor, Salma Sultana, Francis J. Emmatty, and Vinay V. Panicker

1 Introduction

Cleaning is an essential activity to maintain hygienic surroundings. The core activity of cleaning is sweeping, which is the removal of dirt, dust and other extraneous substances from surfaces (Kumar and Kumar 2008). It has an important role to play in the conservation of public health and hygiene. A clean work environment also promotes the quality and efficiency of surroundings, whereas a dirty environment leads to accidents and increase the risk of exposure to various irritants. Hence a country-wide campaign, 'Swachh Bharat Abhiyan' is being started in India to ensure cleaning streets, roads and infrastructure in cities, towns and rural areas.

The street sweepers play a crucial role in the cleaning activities; nevertheless, their work exposes them to a variety of occupational risks like respiratory troubles, musculoskeletal disorders (MSD), eye and skin diseases, and accidents. In this context, the occupational health of sweepers needs to be researched. Sweeping is often an occupation that requires more physical labour and is often characterized by the high occurrence of musculoskeletal disorders (MSD). The major ergonomic risk factors that contribute to musculoskeletal conditions are uncomfortable working postures; substantial force applications; repeated moments; lifting loads; and inappropriate work equipment design (Jeong 2016; Charles et al. 2009).

This article addresses the appropriateness of the types of sweeping activities which are usually carried out. Three varying lengths of broom for three digital manikins at 5th, 50th and 95th percentile are considered. The types of sweeping activities include sweeping with a broom in a single hand, one broom in each hand, and a broom with an extended handle. It was observed that a few of the workers are using two brooms, one

---

A. A. Vishnu · K. S. Siddaveeraiah · K. V. Rahul · S. Nirmal · P. Thalor · S. Sultana · F. J. Emmatty · V. V. Panicker (✉)
Industrial Engineering Laboratory, Department of Mechanical Engineering, National Institute of Technology Calicut, Calicut, Kerala, India
e-mail: vinay@nitc.ac.in

© The Author(s), under exclusive license to Springer Nature Singapore Pte Ltd. 2022, corrected publication 2022
D. Bijulal et al. (eds.), *Technology-Enabled Work-System Design*, Design Science and Innovation, https://doi.org/10.1007/978-981-16-1884-0_8

in each hand for the sweeping task. Hence this method was included in the analysis. Digital human modelling is done using Jack simulation software. Compressive forces on L4/L5 intervertebral disc for the critical postures identified in Rapid Upper Limb Assessment (RULA) are considered for statistical analysis.

After the introduction, the remaining sections of the paper are organized as follows: In the second section, the literature review is covered, addressing the general theoretical context for this research, including occupational risks, musculoskeletal disorders and digital human modelling. The third section presents the problem description and the fourth section describes the methodology. In the fifth section, the results of various sweeping activities are analysed, and the findings are discussed in detail. The paper concludes in the sixth section.

2 Review of Literature

A worker in cleaning activity executes several repeated motions every day, including those that require significant forces to be applied. Street sweeping requires the workers to perform tasks with awkward postures and repetitive movements that can inflict discomfort. Unnatural postures can result in high energy expenditure demands and high work stress on cleaners that can lead to MSDs such as upper limb disorders, lower limb disorders and less work satisfaction that can influence the work productivity and quality (Jeong 2016). Occupational risks include infections, respiratory problems, skin allergies, psychological and musculoskeletal disorders (Charles et al. 2009).

Janitorial work ranked the 22nd highest potential profession among males for back pain and the fourth-highest among females. Among the articles published based on occupational hazards between 1981 and 2005, only five studies considered musculoskeletal disorders as a potential cause for the health outcomes of workers (Charles et al. 2009). The postural risk associated with the cleaning work is not given as much importance as other infections, and air-borne diseases were given.

Various articles on musculoskeletal risk factors are published, while only a few studies are concerned with the design of equipment, the work environment and conditions influencing sweepers. There is a need for research on sweeping tools and individual risk factors (Kumar and Kumar 2008).

Musculoskeletal problems are analysed and compared between different ways of cleaning with appropriate cleaning equipment, experimentally such as mopping method versus scrub and cloth method (Søgaard et al. 1996), sweeping with an extended broom and a traditional broom (Kumar 2002). Ergonomic tool design is critical when it comes to a physical activity that requires a certain amount of time for completion.

The issues with the application and usage of ergonomic tools and techniques in the mopping method were highlighted and addressed in a report by Ohrling et al. (2012). In the case of sweeping, Kumar (2002) has researched tool design, but the factors such as anthropometry and different sizes of the broom are neglected. Anthropometry is the least considered factor while studying musculoskeletal disorders. Only a few studies have considered the anthropometry of the population of cleaners while other studies (Heliövaara 1987; Nathan et al. 1993) took a population other than cleaners for analysis (Kumar and Kumar 2008).

For an ergonomic evaluation of sweeping task, Kumar (2004) (Kumar 2002) adopted a methodology of digital human modelling (DHM) in Jack simulation software and measured the L4/L5 compressive force at spine during the work activity. DHM is used for the proactive ergonomic research activity and assessment of the workspace and its design/redesign. DHM tools enable a designer to create an avatar (virtual human) with different population percentiles in a computer for improving the physical aspects of the work system (Tripathi et al. 2014). While DHM has been extensively used in manufacturing and industrial applications for ergonomic and safety research, it is now time to use it for small-scale applications such as sweeping and cleaning.

3 Problem Description

Identifying the awkward postures which lead to musculoskeletal disorders requires a systematic analysis of sweeping activity. Hence, the sweeping task has been observed, and a digital human model is developed using Jack simulation software. Three different scenarios with a broom in a single hand, one broom in each hand, and a broom with an extended handle are considered for assessment of postures with different digital manikins at the 5th, 50th and 95th percentile according to the ASIAN_INDIAN_NID97 database in anthropometric scaling.

Even though brooms with extended handles are available, many sweepers use the traditional broom as it is more efficient in cleaning but do not recognize the fact about musculoskeletal disorders. Therefore, the tool design should consider several factors to verify whether they are suitable ergonomically. In this work, normal brooms of sizes 75 cm, 100 cm and 125 cm weighing 320 g, 400 g and 480 g, respectively, are used in the different combinations of anthropometric digital human models. The length is 180 cm in the case of a broom with an extended handle weighing 1 kg. Hence, in total, there are nine cases each in the first two scenarios for analysis, and three cases in the third scenario for a broom with an extended handle. The L4/L5 compressive force in the spine is the response variable in each scenario.

4 Methodology

Based on the observation of different sweepers, an average sweeper sweeps continuously for about 6 s and the cycle repeats. Initially, a single broom and an extended handle broom are made using SolidWorks software. Brooms are imported and 21 digital human models are developed accordingly contemplating all the above details

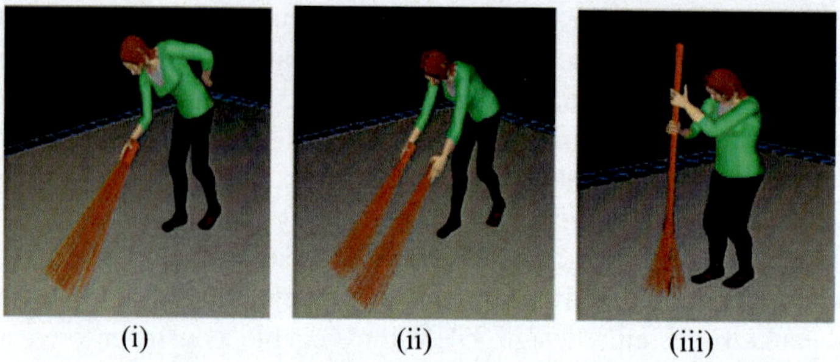

Fig. 1 **i** Digital human model for single broom analysis. **ii** Digital human model for double broom analysis. **iii** Digital human model for extended broom analysis

in Jack software as depicted in Fig. 1. While developing the model, a force of 5 N is given at the start of each step on the right hand of digital manikin for a single broom and for a broom with extended handle analysis, a force of 5 N on right hand and a force of 3 N on left hand of digital manikin for double broom analysis assuming the worker to be a right-handed person and care is taken to produce an exact imitation of real-time workers in the simulation model.

Out of the 21, nine simulation models are of single broom analysis, nine models are of double broom analysis which has lengths of broom and human percentiles as varying factors and three models are of the broom with extended handle analysis which has only human percentiles as a varying factor. Two postures having the highest and second highest values of L4/L5 compressive forces are noted with care for all digital human models.

Ten replications are taken for each model. While replicating a digital model, the maximum value of the two postures in every replication is noted down, and the same process is repeated for all 21 models. A statistical analysis using ANOVA with replications is carried out to test the significance of two factors, namely, the length of the broom and the human percentile, on the response variable, i.e. L4/L5 compressive force in each type of broom analysis.

Based on the L4/L5 compressive forces obtained from simulations, two-way ANOVA with 10 replications for the single and double broom and a one-way ANOVA with 10 replications for a broom with extended handle are done using MS-EXCEL at a level of significance $\alpha = 0.05$. The following hypotheses are formulated for the analysis:

H_0: There is no significant influence of the factors on the response variable.

H_1: There is a significant influence of the factors on the response variable.

In two-way ANOVA, the two factors considered are the length of broom and percentile of DHM. In the one-way ANOVA, the factor considered is the percentile of DHM.

5 Results and Discussions

The values of L4/L5 compressive forces of single broom and double broom analyses are found almost similar as shown in Figs. 2 and 3, for all lengths of broom and percentiles of DHM. This finding can be because of the similar postures needed for a sweeping task using a single broom and a double broom.

In the scenario of sweeping using the single broom and double broom, the L4/L5 compressive force is found to increase with a decrease in the length of the broom, irrespective of the DHM percentile. The highest L4/L5 compressive force is observed in the 95th percentile of DHM, and the least is observed in the 5th percentile of DHM. The 50th percentile of DHM is an intermediate of both for all three lengths of the broom. For all three percentiles of DHM, the broom with the maximum length (125 cm) gave the minimum L4/L5 compressive force values by 15%.

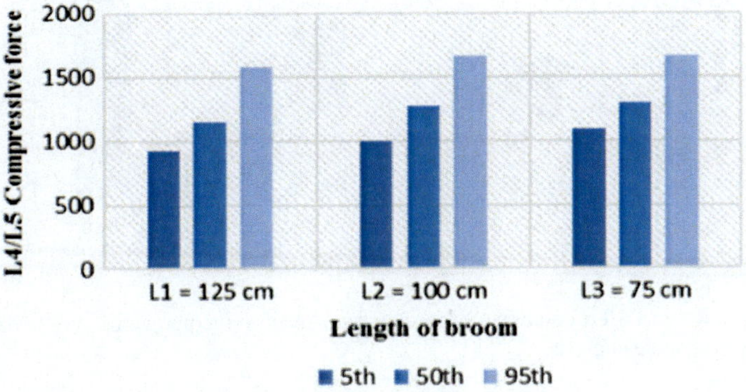

Fig. 2 Variation of L4/L5 compressive force with respect to the length of broom and percentile of DHM for single broom analysis

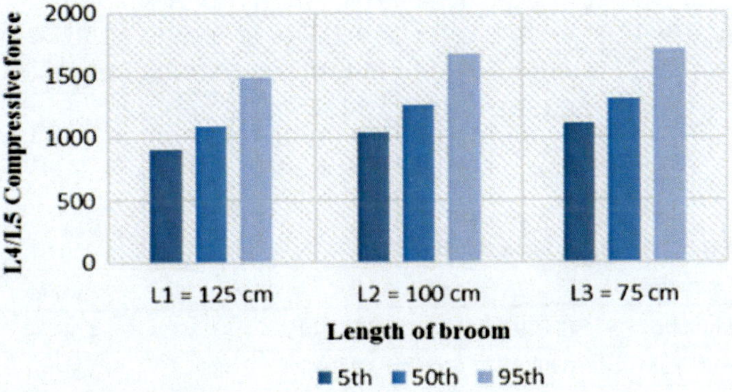

Fig. 3 Variation of L4/L5 compressive force with respect to the length of broom and percentile of DHM for double broom analysis

Using the broom with an extended handle for sweeping, it is observed that the 95th percentile of DHM has the highest L4/L5 compressive force followed by the 50th percentile and the 5th percentile has the least as shown in Fig. 4.

From the above analyses (see Tables 1 and 2) as P-value < 0.05, the null hypothesis, i.e. H_0 is rejected and is concluded that there is a significant influence on L4/L5 compressive force in terms of the change in the length of broom and the change in percentile of DHM. Also, there is a significant interaction between these two factors.

From the above analysis (see Table 3) as P-value < 0.05, the null hypothesis, i.e. H_0 is rejected and is concluded that there is a significant influence on the L4/L5 compressive forces in terms of the change in percentile of DHM.

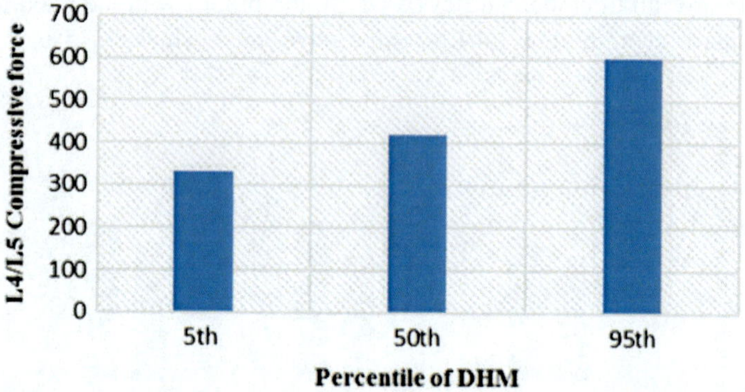

Fig. 4 Variation of L4/L5 compressive force with respect to the percentile of DHM for a broom with extended handle analysis

Table 1 Two-way ANOVA for single broom

Source of variation	*SS*	*df*	*MS*	*F*	*P-value*	*F crit*
Length of broom	433,420.6	2	216,710	171.84	7.201E-30	3.109
Percentile of DHM	6,044,256	2	3,022,128	2396.4	8.617E-73	3.109
Interaction	61,636.8	4	15,409.2	12.219	8.065E-08	2.484
Error	102,152.1	81	1261.14			
Total	6,641,465	89				

Table 2 Two-way ANOVA for double broom

Source of variation	*SS*	*df*	*MS*	*F*	*P-value*	*F crit*
Length of broom	548,879.6	2	274,440	318.33	4.3E-39	3.109
Percentile of DHM	6,699,088	2	3,349,544	3885.3	3.5E-81	3.109
Interaction	50,696.21	4	12,674.1	14.701	4.5E-09	2.484
Error	69,831.11	81	862.112			
Total	7,368,494	89				

Table 3 One-way ANOVA for a broom with an extended handle

Source of variation	*SS*	*df*	*MS*	*F*	*P-value*	*F crit*
Percentiles of DHM	587,435	2	293,717.5	296.9	4.16E-19	3.35
Error	26,713.53	27	989.38985			
Total	614,148.5	29				

Multiple comparison tests are carried out in each scenario for every factor to know whether all the levels in a factor are significantly different. It is observed that all levels are significantly influencing the response variable in all three scenarios.

6 Conclusions

In the case of normal brooms, the broom with the maximum length is recommended, as the compressive forces at L4/L5 are lesser. A broom with an extended handle, when compared to normal brooms, gave the lesser L4/L5 compressive force values for all three percentiles of DHM. But practically workers preferred to use normal brooms because broom with extended handle was less efficient in cleaning as compared to the normal broom. The surface area swept by a normal broom can be varied according to the requirement of the task when compared with the broom with the extended handle.

The use of the double broom for the sweeping task is found to be preferable over a single broom, considering all percentiles of humans, as the compressive force at L4/L5 are lesser. The compressive force at L4/L5 is lesser for all percentiles due to the similarity in postures encountered during sweeping activity, as cleaning with a double broom has an additional advantage of efficient cleaning due to the use of both hands when compared with the single broom.

The present study has considered a digital human modelling approach for the assessment of sweeping tasks under various scenarios. The findings of the research can be further validated by physiological measurements like heart rate and energy expenditure incurred by the workers.

References

Charles LE, Loomis D, Demissie Z (2009) Occupational hazards experienced by cleaning workers and janitors: a review of the epidemiologic literature. Work 34(1):105–116

Heliövaara M (1987) Occupation and risk of herniated lumbar intervertebral disc or sciatica leading to hospitalization. J Chronic Dis 40(3):259–264

Jeong BY (2016) Occupational deaths and injuries by the types of street cleaning process. Int J Occup Saf Ergon 23(1):76–82

Kumar R (2002) Ergonomic assessment of two different cleaning tools while cleaning floor. Int Cyber Conf Ergon

Kumar R, Kumar S (2008) Musculoskeletal risk factors in cleaning occupation—a literature review. Int J Ind Ergon 38(2):158–170

Nathan PA, Keniston RC, Meadows KD, Lockwood RS (1993) Validation of occupational hand use categories. J Occup Med 35(10):1034–1042

Öhrling T, Kumar R, Abrahamsson L (2012) Assessment of the development and implementation of tools in contract cleaning. Appl Ergon 43(4):687–694

Søgaard K, Fallentin N, Nielsen J (1996) Workload during floor cleaning. The effect of cleaning methods and work technique. Eur J Appl Physiol Occup Physiol 73(2):73–81

Tripathi B, Rajesh R, Maiti J (2014) Ergonomic evaluation of billet mould maintenance using hierarchical task analysis, biomechanical modeling and digital human modeling. Comput Aid Des Appl 12(3):256–269

Chapter 9
Working Conditions of Indian Glass Artware Workers and Association with Musculoskeletal Symptoms

Bhawana Rathore, Ashok K. Pundir, and Rauf Iqbal

1 Introduction

Handicraft industry is regarded as highly labor-intensive in many developing countries (Das et al. 2018). Handicraft industries are those industries where the machine does not dominate and is much decentralized. India is one of the culturally rich countries in the world and is considered a leading hub for employment in India (Meena et al. 2014). In the handicraft sector, skilled labor resources are considered primary input and the main production factors (Mahdavi et al. 2018). However, workers suffered from various symptoms and injuries like musculoskeletal disorders (Sain et al. 2018). Work-related musculoskeletal disorders (WMSDs) are one of the main causes of productivity losses, working time losses, and employee injuries around the world. Musculoskeletal diseases cause worker's compensation, medical expense, and sick leaves (Meerding et al. 2005; Ijzelenberg and Burdorf 1976). Thus, control and reduction of WMSDs among the worker represent a major ergonomics issue. Handicraft industries are considered as an artistic activity where useful and decorative objects are made traditionally with hands or by the use of a simple tool. The handicraft industry includes several occupations, but some operations and activities share some common work characteristics such as prolonged sitting, repetitive tasks, and forceful movement, as shown in Fig. 1. Indian glass artware industry is considered as one of the types of occupations among handicraft industries.

B. Rathore (✉) · A. K. Pundir · R. Iqbal
National Institute of Industrial Engineering Mumbai, Mumbai 400087, India
e-mail: Bhawana.Rathore.2017@nitie.ac.in

© The Author(s), under exclusive license to Springer Nature Singapore Pte Ltd. 2022, corrected publication 2022

D. Bijulal et al. (eds.), *Technology-Enabled Work-System Design*, Design Science and Innovation, https://doi.org/10.1007/978-981-16-1884-0_9

Fig. 1 Working posture adopted to perform an activity

2 Materials and Methods

A cross-sectional study was conducted in the glass artware industry of the labor-intensive handicraft sector. The study was carried out in the Firozabad district of Uttar Pradesh, India, which is also called a glass city. A total of 38 male workers ranging from 25 to 66 years participated. A field survey has been conducted for problem identification of musculoskeletal disorders in body regions of the neck, lower back, upper back, shoulders, and wrist/hands associated with glass artware. Nordic questionnaire (Kuorinka et al. 1987) was used to measure the prevalence of musculoskeletal disorders and extremity in their body regions among glass artware workers. The data has been analyzed using SPSS software version 16.

3 Result

The existing workstations used by the artware workers have been analyzed in detail. Around 40% of a cluster of glass bangles in Firozabad is unorganized, and they are not registered and operating as small-scale industries. All the workers in the production system were male and engaged in continuously performing their work for 8–9 h per day with a break of 30–60 min during the entire day of work in the factory. To avoid selection bias, all 38 workers were selected randomly from the Firozabad district of Uttar Pradesh. Improper design of a workstation, awkward posture, excess heat stress, inadequate lighting, and improper ventilation provide poor working conditions to the workers. About 90% of the work is performed manually, and one machine is used to give only a spiral shape to the bangles.

Table 1 Demographic profile of participants

Parameters (variables)	Average	SD	Range
Age (years)	40.13	9.73	25–66
Height (cm)	168.98	5.26	158.50–179.83
Work experience (years)	17.47	7.85	7–40

Table 2 Distribution of participants in different age groups

Explanatory variables	N	%
Male	38	100
Age (in years) range		
25–30	7	18.42
31–35	6	15.79
36–40	7	18.42
41–45	11	28.95
46–50	1	2.63
51–55	1	2.63
56–60	4	10.53
61–66	1	2.63

3.1 Socio-Demographic Characteristics of the Study

In this study, 38 male workers (N = 38) subjects participated. The average age was 40.13 years; average height 168.98 cm; and mean weight being 67.29 as presented in Tables 1 and 2.

4 Discussion

The results of the study found that glass artware workers in India are engaged in a repetitive and prolonged sitting task. The finding showed that 90.5% of the workers covered in the study had faced a musculoskeletal problem in at least one body area, as shown in Table 3. The study carried by Choobineh et al. (2004) on carpet mending operation has reported that shoulders and lower back pain were significantly more prevalent compared to other body regions. Similar findings carried out by Dianat and Karimi (2016) on handicraft workers handling sewing activities have reported that 76.2% of the respondents report symptoms. Therefore, this study can be included as a job with a high risk level of WMSDs.

In addition, lower back, and both shoulders and wrists had experienced high musculoskeletal pain relative to other body parts, as shown in Fig. 2. There is a high demand for the implementation of ergonomics intervention measures in the glass artware industry. This study also revealed that around 60% of the population is

Table 3 Prevalence of MSD symptoms in different body regions of glass artware workers

MSD reported in body regions	N	%
Neck	8	10.81
Both shoulders	10	13.51
Right shoulder	4	5.41
Left shoulder	4	5.41
Lower back	14	18.92
Upper back	4	5.41
Left wrist	5	6.76
Right wrist	5	6.76
Both wrists	14	18.92
Both elbows	1	1.35
Left elbow	0	0.00
Knees	5	6.76
Ankles/feet	0	0.00
Hips/thighs/buttocks	0	0.00
Right elbow	0	0.00

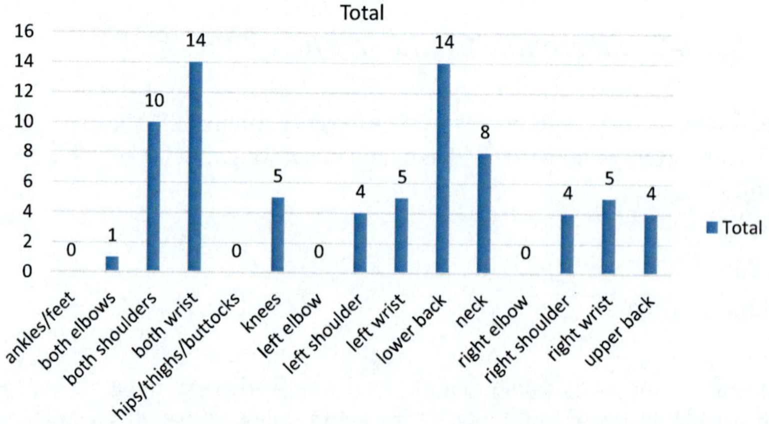

Fig. 2 Graphically presentation of the prevalence of MSD symptoms

suffering from severe pain in different body parts due to musculoskeletal symptoms, as shown in Fig. 3.

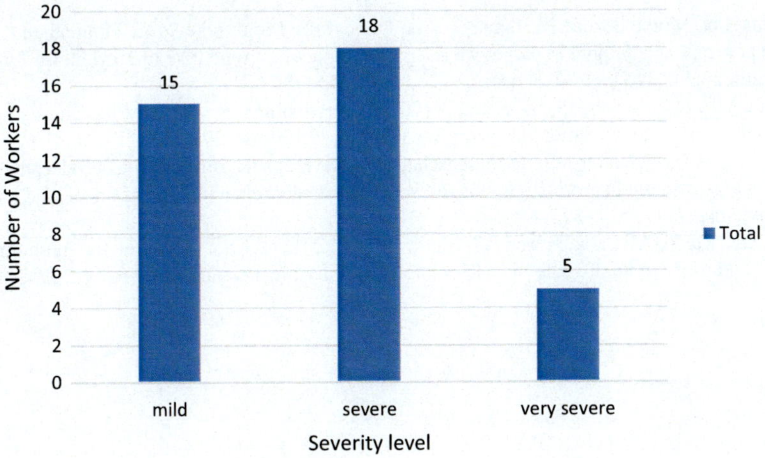

Fig. 3 Severity level in glass artware workers

5 Conclusion

This study showed that the musculoskeletal health of the workers is significantly impacted, and there is a high chance of risk that the workers would result in WMSDs in their different body regions. Most of the workers in this study are suffering from lower back, and both shoulders and wrist problems because of the monotonous and repetitive nature of the job and the poor workstation design. There is a loss of strength in the hands/wrist because of the prolonged duration of work. The NMQ also suggests the statistics of the symptoms and severity in different body parts of the worker. The findings of this study suggest that there is a need for ergonomic intervention on the redesign of the workstation and rectifying the working posture of the artware workers.

References

Choobineh A, Tosian R, Alhamdi Z, Davarzanie M (2004) Ergonomic intervention in carpet mending operation. Appl Ergon 35(5):493–496

Das D, Kumar A, Sharma M (2018) A Systematic review of work-related musculoskeletal disorders among handicraft workers. Int J Occup Saf Ergon 1–30

Dianat I, Karimi MA (2016) Musculoskeletal symptoms among handicraft workers engaged in hand sewing tasks. J Occup Health 58(6):644–652

Ijzelenberg W, Burdorf A (2005) Risk factors for musculoskeletal symptoms and ensuing health care use and sick leave. Spine (Phila. Pa. 1976) 30(13):1550–1556

Kuorinka I et al (1987) Standardised nordic questionnaires for the analysis of musculoskeletal symptoms. Appl Ergon 18(3):233–237

Mahdavi N, Motamedzade M, Jamshidi AA, Darvishi E, Moghimbeygi A, Moghadam RH (2018) Upper trapezius fatigue in carpet weaving: the impact of a repetitive task cycle. Int J Occup Saf Ergon 24(1):41–51

Meena ML, Dangayach GS, Bhardwaj A (2014) Investigating ergonomic issues among workers in hand block textile printing industries. Int J Bus Syst Res 8(4):392

Meerding WJ, IJzelenberg W, Koopmanschap MA, Severens JL, Burdorf A (2005) Health problems lead to considerable productivity loss at work among workers with high physical load jobs. J Clin Epidemiol 58(5):517–523

Sain MK, Meena ML, Dangayach GS, Bhardwaj AK (2018) A brief procedure for ergonomic design of hand tool for small-scale industries hand tool musculoskeletal disorders, pp 209–214

Chapter 10
Ergonomic Design of a Troweling Tool

Richard Laurence, G. Kiran, Solomon Darius Gnanaraj, and Saleem Ahmed

1 Introduction

Flooring is the general term for a permanent covering of a floor, or for the work of installing such a floor covering. Floor covering is a term to describe any finishing material applied over a floor structure to provide a walking surface. The most common flooring done in buildings is solid ground flooring. But before flooring is done on the floor, it is blinded. Concrete blinding is the process of pouring a thin layer of concrete over the floor of a building. The purpose of this covering is to seal in all underlying material and prevent dirt and mud from interfering with the structure. The concrete blinding makes up the bottom layer of the floor slab in a house and creates strong floors (Fig. 1).

After blinding, the floor surface is leveled, flattened and smoothened by using a concrete finishing trowel, as shown in Fig. 2. This method is called troweling. Troweling is a method that produces a hard, smooth, dense surface and should be done immediately after floating. Troweling can be done by machine or by hand. A float trowel or finishing trowel is used for troweling which is rectangular in shape. In addition to removing surface imperfections, the trowel will compact the concrete and prepare it for further processes. A trowel can be a small hand tool or a large bull float with a long handle, or a power trowel with an engine. The problems that occur in smoothing the surface for the workers are: while flattening, the worker has to squat and work which causes back pain and discomfort in many parts of the body, due to the awkward posture as shown in Fig. 3. This paper reports an ergonomically designed troweling tool that will reduce the physical discomfort.

Information was collected through the internet to find the troweling tools available in the market. Details are shown in Table 1.

R. Laurence · G. Kiran · S. D. Gnanaraj (✉) · S. Ahmed
Vellore Institute of Technology (VIT), Vellore 632014, India
e-mail: dariusgnanaraj.s@vit.ac.in

© The Author(s), under exclusive license to Springer Nature Singapore Pte Ltd. 2022,
corrected publication 2022

D. Bijulal et al. (eds.), *Technology-Enabled Work-System Design*, Design Science and Innovation, https://doi.org/10.1007/978-981-16-1884-0_10

Fig. 1 **a)** Bottom layer of the floor **b)** Concrete blinding of the floor

Fig. 2 A Troweling Tool (https://www.indiamart.com/proddetail/round-end-finishing-trowel-with-long-shank-9476015197.html)

Fig. 3 Troweling process

Table 1 Troweling tools available in the market

Product	Product ranufacturer	Product description	Price (in rupees)
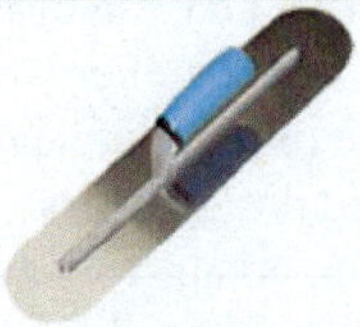	Bonway Construction Products Pvt Ltd (https://www.indiamart.com/proddetail/round-end-finishing-trowel-with-long-shank-9476015197.html)	Bonway round end finishing trowel with long shank	1480/piece
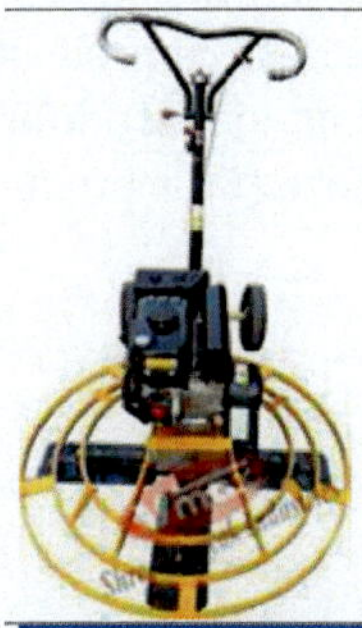	Shree Vinayak Industries (https://shreevinayakind.com/index.php/product/concrete-power-trowel-svf-1000pi/)	Features: This power trowel is a high effective floor finishing machine	50,000/piece
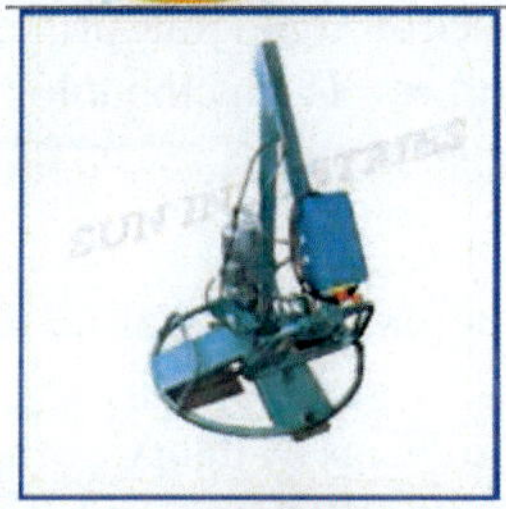	Sun Industries (http://www.sunind.co.in/power-trowel-floater-4741763.html)	The product range includes a wide range of trimix system,	43,000/piece
	Ride On Enterprises (https://www.google.com/search?q=ride+on+enterprise+ST+Trunks+Ride+On+Trowel+Machine+has+a+capacity+of+35+ton.&client=firefox-b-d&tbm=isch&source=iu&ictx=1&fir=TEDQqKlZDyQ7lM%253A%252C2GScEdBihCcRMM%252C_&vet=1&usg=AI4_-kQP7bbsyxPjyLTs5SD0atiVIJkkkA&sa=X&ved=2ahUKEwjMhOS_o4voAhXn4zgGHXg0DBYQ9QEwA3oECAoQBQ#imgrc=TEDQqKlZDyQ7lM)	ST trunks ride on trowel machine has a capacity of 35 ton	3.95 lakhs/unit

2 Literature Review

Bhatia et al. (Bhatia et al. 2016) found the dimensions of the optimal handle of the trowel using anthropometric data collected from 40 participants. EMG studies were carried out and found that the muscle fatigue was less in the case of ergonomically designed trowel handles compared to the trowels made of standard designs. Strasser et al. (1996) evaluated an ergonomically designed handle in comparison with two standard types, both with a round cross-section of the handles and either a straight neck or a swan's neck. The ergonomically designed handle enabled a specific relief of the strain in the grip musculature and the ulnar deviation muscles. Bisht and Khan (2013) recognized an important issue in the design of hand-held industrial products to identify the factors that lead to human comfort and those leading to discomfort. Their work provides a solid foundation on which any future research for product development and assessment can be performed and analyzed.

3 Objective of the Study and Methodology

The objective of the study is to design a new troweling tool based on ergonomic principles to reduce the discomfort experienced by masons and workers. The methodology is given below:

- Literature review
- Collection of information about trowels available in the market as well as in workplaces
- Collection of information about the occupational health problems by interviewing masons and workers
- Ideation of new designs based on the principles of ergonomics
- Fabrication of a prototype based on the best design
- Field test using the prototype and improvement of the prototype based on the feedback given by users

4 Field Study and User's Feedback

Finishing of the concrete surface is done after compaction to obtain a uniform and smooth surface. There are various steps involved in the finishing as shown below:

(a) Screed: This is the process of striking excess concrete using a straight edge (a plank). The surface is struck off by moving the straight edge back and forth across the top of the surface. This removes the bumps and holes and gives a uniform leveled surface.

(b) Floating: It removes the irregularities left after the screed and firmly embeds the large aggregate particles. This is done using wooden floats which moved forward and backward.

(c) Troweling: It is done to obtain a very smooth and highly wear-resistant surface. It is done using a steel trowel. Troweling should not be done on the surface that is not subjected to floating. Spreading of dry cement on wet surfaces should be avoided as it causes weaker surfaces cracks and blisters.

To study the procedures followed during construction, a construction site was visited which is situated in VG Rao Nagar, Katpadi, Vellore. During the visit, the site workers were interviewed to get feedback on the troweling tool they were using as shown in Figs. 4, 5, 6 and 7.

The feedback received from the contractor, mason and workers is given below:

Fig. 4 A conversation with a contractor

Fig. 5 A conversation with workers

Fig. 6 A conversation with a mason

Fig. 7 A conversation with a mason

- Concrete is poured and a plank is normally used to remove extra concrete and to level the floor. A trowel is used to finish the surface by making it smooth.
- A manual trowel is used for smaller areas and power trowels are used when the area is large.
- They normally squat and work and claim that only if they squat, they can do the finishing work properly.
- Due to squatting, they experience back pain after working for long hours.

Figure 8 shows a hand trowel used in the field, which is 8 inches long and 4 inches wide. The blade is made of hardened tempered steel and the handle is made of plastic. The price is just Rs. 90 and it is available in the market. The design of this tool is good but not ergonomic. The edges of the trowel are sharp; it might hurt the worker. During troweling if mason wants to change the direction of his hand motion he has to move the trowel with his wrist. The size of the trowel is small, so for a larger area

Fig. 8 A hand trowel

it is not useful. More importantly, the worker has to squat and work. With this trowel the work efficiency is less and occupational health hazards are more.

5 Design Ideation and Model

Based on the user's feedback, market research and review of the hand trowel, some ideation sketches having two trowels were made. One trowel is for roughing work and the follower trowel is provided for finishing work.

5.1 Ieation One

The ideation sketch of concrete finishing trowel as shown in Fig. 9 solves the ergonomic issues related to back pain. It consists of two steel trowels (Trowel 1 and Trowel 2), each one having curved edges. There is a rectangular platform made of wood holding trowel 1. A handle made of steel that can be elongated as per the height of the mason is attached to the wooden platform. A connecting rod is used to connect the trowel 2. The mason can occupy a normal posture and easily work in this setup and solve the problem of back pain.

To make this idea work, the ideas were sketched and a cardboard model was made as shown in Fig. 10 to understand the practical difficulties while working with this model.

The mock-up was tried and moved in all ways and directions. But while trying, it couldn't completely finish troweling the corners. At the corners the trowel 1 can only be used and not trowel 2.

Limits of ideation one design are:

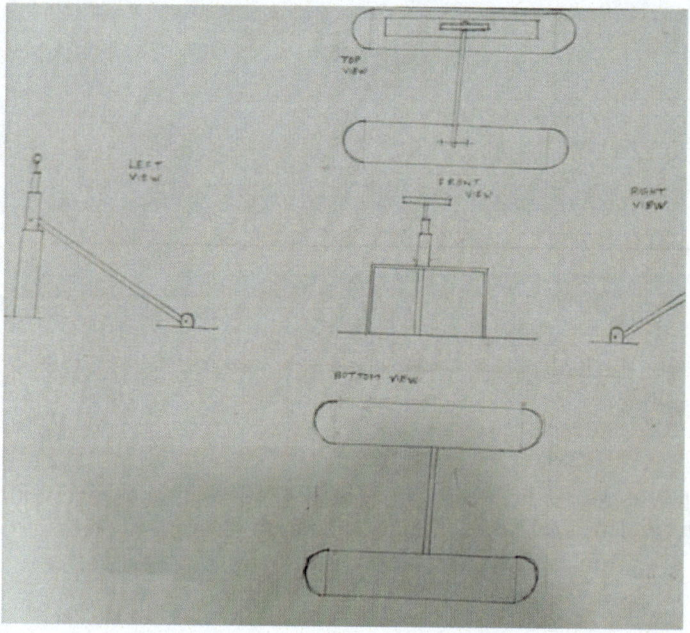

Fig. 9 Ideation sketch of a trowel

Fig. 10 Cardboard model of a trowel

- Making the platform with wood and handle with steel makes the equipment heavy which is not very comfortable.
- Connecting the connecting rod to trowel 1 was difficult.

5.2 Ideation Two

The troweling tool made of wood and steel rod has more weight. The ideation sketch of this (Fig. 11) concrete finishing tool is simple. It has two blades slightly folded to avoid sharp pointed edges. The platform is avoided to reduce weight, and the handles and connecting rod will be of lighter material. By using the 3D software such as Alias and keyshot (for rendering), the 3D model of the equipment was designed digitally to give the idea how the product will look like (Figs. 12 and 13).

The final prototype is fabricated (Fig. 14). The troweling blades are made of

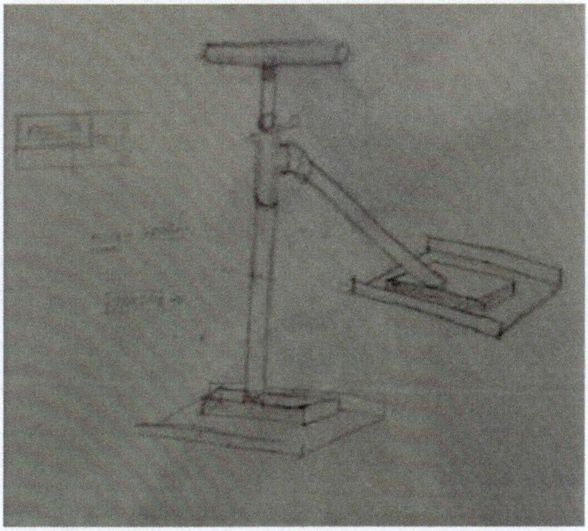

Fig. 11 Design ideation

Fig. 12 3D model

Fig. 13 3D model

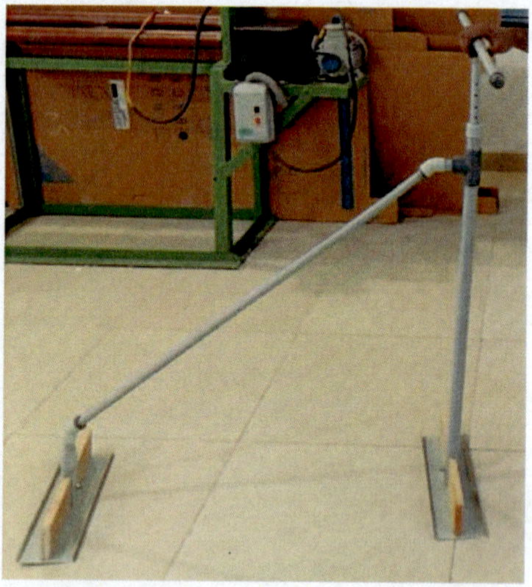

Fig. 14 Low-cost model

galvanized steel, upon which the wood is fixed using screws on both the blades having dimensions of 16 × 4.5 inches. The connecting rod and the handle are made of PVC pipes as they are light in weight. The height of the handle can be adjusted as per the elbow height of workers. The prototype is made considering anthropometric data collected from 20 workers by measuring their elbow height measured in standing posture.

6 Results and Discussions

Rapid entire body assessment (REBA) was carried out on postures occupied by persons while using the traditional troweling tool (Fig. 15) and an ergonomically designed model of a troweling tool (Fig. 16).

Fig. 15 A person using a traditional troweling tool

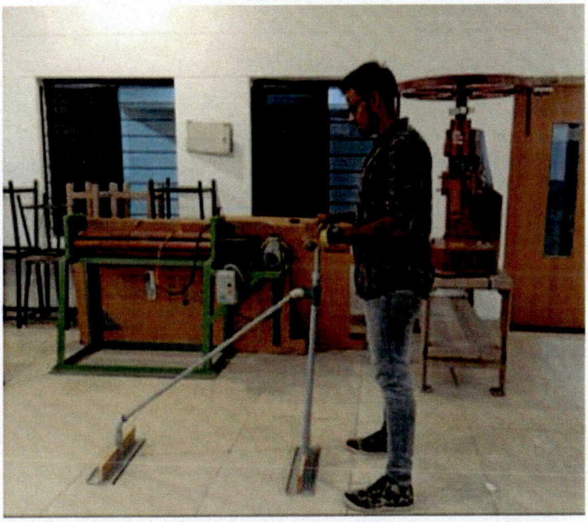

Fig. 16 Posture of a person testing an ergonomically designed model of a troweling tool

The result shown in the first case is alarming as the REBA score is 8, which means high risk and immediate investigation is needed and change should be implemented. The results shown in the second case are excellent as the REBA score is 3, which means low risk and change may be needed if necessary. It is proposed to make an actual troweling tool using aluminum parts and try it in construction sites to know the problems experienced by workers so that the design can be improved further. Even though expensive power troweling machines are available as shown in Table 1, and if a low-cost hand-operated troweling tool is developed, the workers will be able to purchase them, use them and work in the natural posture, leading to a reduction in occupational health hazards.

7 Conclusion

A low-cost model of an ergonomically designed troweling tool is presented. REBA score was found to decrease from 8 to 3 when the posture changed from squatting to standing by using the newly designed troweling tool. Even though there are expensive power tools available in the market, this low-cost model will help many masons and workers to have comfortable work postures to improve their occupational health.

Acknowledgements The authors thank the management and administration of VIT Vellore for providing support and cooperation for carrying out this research work. The masons and workers at the construction sites at VG Rao Nagar, Katpadi, Vellore, Tamil Nadu, India are thanked for sharing their time and experience in giving information about the problems available with existing trowels and also suggestions to improve the prototype of the newly designed trowel based on the principles of ergonomics.

References

Bhatia A, Kalra P, Singh J (2016) Ergonomic Design of trowel handle. Int Res J Eng Tech (IRJET) 3(7):819–824, ISSN: 2395–0056

Bisht DS, Khan MR (2013) Ergonomics assessment methods for the evaluation of hand held industrial products: a review. In: Proceedings of the world congress on engineering 2013, vol I. London, UK, 3–5 July 2013

Strasser H, Wang B, Hoffmann A (1996) EMG and subject evaluation of hand tools: the example of mason's trowel. Int J Ind Ergon 18(1):91–106

https://www.hoganco.com/product-category/construction-products/hand-trowels/

https://www.indiamart.com/proddetail/round-end-finishing-trowel-with-long-shank-9476015197.html

https://shreevinayakind.com/index.php/product/concrete-power-trowel-svf-1000pi/

http://www.sunind.co.in/power-trowel-floater-4741763.html

https://www.google.com/search?q=ride+on+enterprise+ST+Trunks+Ride+On+Trowel+Machine+has+a+capacity+of+35+ton.&client=firefox-b-d&tbm=isch&source=iu&ictx=1&fir=TEDQqKlZDyQ7lM%253A%252C2GScEdBihCcRMM%252C_&vet=1&usg=AI4_-kQP7bbsyxPjyLTs5SD0atiVIJkkkA&sa=X&ved=2ahUKEwjMhOS_o4voAhXn4zgGHXg0DBYQ9QEwA3oECAoQBQ#imgrc=TEDQqKlZDyQ7lM

Chapter 11
Variation of Hand Strength Among Bike Riders After Experimental Trials

K. M. Jishnu Devadas and S. A. Binoosh

1 Introduction

Two-wheelers are one of the popular vehicles for city transport in India and their usage has increased tremendously during the present decade. Reports indicate that the two-wheelers registration in India has increased to 130 million in the year 2013 (https://data.gov.in/catalog/total-number-registered-motor-vehicles-india). The number of road accidents also indicates an increasing trend during this period (http://pibphoto.nic.in/documents/rlink/2016/jun/p20166905.pdf). In the case of two-wheeler riding, one of the major reasons for road accidents is driver fatigue (Dagli 2016). While riding the bike, the rider uses both hands to balance the handle and operate various hand controls. The various hand operations cause muscle fatigue in the palm, fingers, wrist and forearm of the riders. Muscle fatigue is the reduction in capacity to generate maximal force or power output (Vollestad 1997).

The maximum voluntary force of the hand produced by the contraction of the muscles indicates the hand capacity or hand strength (Kong et al. 2011). The hand muscles are subdivided into extrinsic and intrinsic muscles (https://www.revolvy.com/page/Muscles-of-the-hand). The combined contraction of the extrinsic and intrinsic muscles produces the force. There are studies conducted to analyze normal hand conditions and variations by measuring grip strength with a hand dynamometer (Swanson et al. 1970). To analyze the hand condition four hand strength tests were used in the literature: grip, palmar pinch, tip pinch and key pinch (Mathiowetz et al. 1984). The relation between muscle fatigue, discomfort and muscle strength was found useful in studying the effect of physical tasks with hands (Schwid et al. 1999). Another study about physical fatigue revealed the importance of evaluation of

K. M. J. Devadas · S. A. Binoosh (✉)
Department of Mechanical Engineering, College of Engineering Trivandrum, Thiruvananthapuram, Kerala, India
e-mail: binoosh@cet.ac.in

© The Author(s), under exclusive license to Springer Nature Singapore Pte Ltd. 2022, corrected publication 2022

D. Bijulal et al. (eds.), *Technology-Enabled Work-System Design*, Design Science and Innovation, https://doi.org/10.1007/978-981-16-1884-0_11

power grip strength-endurance in finding occupational injury (Robertson et al. 1996). Reliable results were also obtained from these studies regarding hand strength.

The hand strength can be measured with a hand dynamometer. Literature works show that Jamar dynamometer and Saehan dynamometers are excellent in producing clinically usable reliable results (Mathiowetz et al. 1984; Reis and Arantes 2011). Both the equipments are made by the same company under different trade names. Studies have shown that the average of three trials could give the highest test–retest reliability. When the highest score of three trials or score from one trial were utilized, lower correlations were shown (Peters et al. 2011). These concepts of hand strength evaluations were used in the analysis and design of hand-held product usage (Hedge et al. 1999; Berolo et al. 2011).

In the case of analyzing hand strength among bike riders, the effect of factors like road conditions and hand gripping conditions can be explored. Hand regions used for operating the controls and handles can have discomfort or pain due to these factors. During bike riding, the rider uses hands to balance the vehicle, activate the clutch, engage the front brake, adjust throttling, and do other activities like using of horn, indicator and other equipments on a bike. These hand activities together with the hand vibration during riding can cause muscle fatigue. In order to face these issues, it needed to study the situation in detail so that countermeasures can be developed.

The hand strength evaluation method was used in the literature for different purposes but never in the case of vehicle riding. The study was conducted based on this research gap. Analyzing the hand strength variation during riding can help to evaluate the muscle fatigue that occurs due to riding. The factors affecting the hand strength variation were also analyzed in this study.

2 Subjects and Methods

The following sections discuss the details of the subjects involved in this study and the procedure adopted. In a qualitative study, 100 subjects participated while in a field study 20 male bike riders were involved.

2.1 Data Collection

Questionnaire survey and field experiments were used for data collection of the study and it is described in the following sections.

2.1.1 Discomfort Survey

The survey was conducted using a standard hand discomfort questionnaire to collect the data of prevalence and type of hand discomfort among regular bike rides (http://

ergo.human.cornell.edu/ahhandmsquest.html). Data were collected from 100 regular male bike riders with selection criteria of age above 20 years but below 40 years. The sample size provides a 95% confidence level with ± 10 precision when assuming maximum variability. The hand discomforts due to bike riding at various regions of the hands were recorded using the questionnaire. The body mass index (BMI), the average riding hour per day and the experience of the subjects in bike riding were also collected.

In the hand discomfort questionnaire, the various regions of hands are defined based on their sensory stimulation and morphology. The regions of the hand are fingers stimulated by the median nerve (A), fingers stimulated by the ulnar nerve (B), thumb (C), palm (D), thenar muscles (E) and hypothenar muscles (F). The data regarding discomforts in these regions based on severity, frequency and interference with the respondent's daily life were collected using the questionnaire. Rating scores were used to identify the most serious problem.

The frequency of discomfort was rated on a five-point scale starting from never to several times every day. The rating scores were 0, 1.5, 3.5, 5 and 10. The discomfort level was rated on a three-point scale as 'slightly uncomfortable', 'moderately uncomfortable' and 'very uncomfortable' with a rating score of 1, 2 and 3, respectively. The interference of hand discomfort in daily work activities was also rated on a three-point scale as 'not at all', 'slightly interfered', 'substantially interfered' with a rating score of 1, 2 and 3, respectively.

The total discomfort score was calculated by multiplying the 'frequency score', 'discomfort score' and 'interference score'. The data collected from the survey were used to evaluate the hand discomfort problems and correlated with average driving hours a day, driving experience and BMI.

2.1.2 Field Experiment

The hand strength variation during bike riding and its relation to the road conditions and gripping conditions of the hand were evaluated using a field experiment. Grip strength and pinch strength measurements were used to evaluate hand strength. For the experiment, a total of 20 healthy subjects with an age range of 20–26 years were selected. The subject's height was ranged from 162 to 184 cm and weight was ranged from 55 to 98 kg with a BMI of 18.5–29.6 kg/m^2. All subjects were right-handed and they did not report any history of musculoskeletal disorders.

The subjects were instructed to ride the bike continuously for 30 min. The four measures of hand strength were measured before and after the ride to evaluate the variation. Each measurement was carried out for both the left hand and right hand. The average value of three replications of measurements was taken as the measure of strength. The studied parameters of road conditions and gripping conditions were assumed in two levels. For road condition, the two levels were 'good road condition' and 'bad road condition'. For gripping condition, the two levels were 'bike riding without using hand gloves' and 'riding the bike using hand gloves'. The experiments were formed based on repeated measure design. The same group of subjects had undergone the trials involving all possible combinations of the parameters in a random order. The experiments were conducted either in the morning time or in the evening, which allows them to have enough relaxation time from daily work activities. Figure 1 shows the riding posture during experiment trials of the subjects. They were instructed to follow erect posture during the trials in order to maintain uniformity in force application for various handle operations. For each subject, the trials were conducted with a time gap of minimum 5 days.

Before measuring the hand strength, the subject's name, gender, age, height and weight data were recorded. Then the subjects were instructed to seat comfortably on a standard chair with a height of 46 cm (Fig. 2). Among the strength measurements, grip strength was measured first, then key pinch, tip pinch and palmar pinch measurements were carried out. For each of the hand strength measurement, the subjects were seated and positioned as per the recommendations of ASHT (Mathiowetz et al. 1984). The scores of three successive trials were recorded for each strength measurement. The grip strength was measured using the SAEHAN hydraulic hand dynamometer as

Fig. 1 Riding posture during the experimental trial

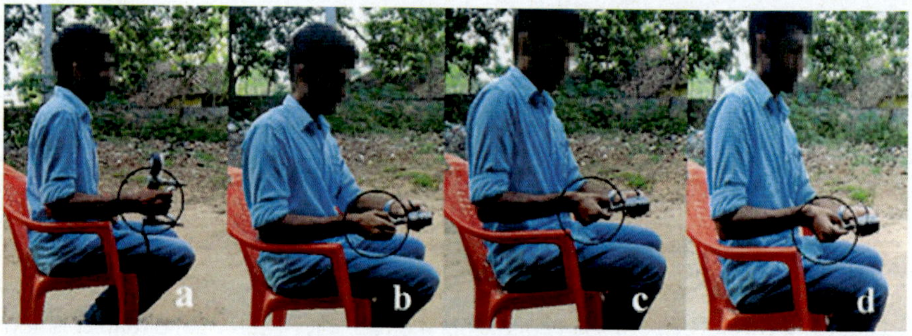

Fig. 2 Measurements of hand strength; **a** grip strength, **b** key pinch strength, **c** tip pinch strength and **d** palmar pinch strength

shown in Fig. 2a. Three measurements were taken before and after the bike riding trials. The hand strength variation is taken as the difference between grip strength before and after bike riding. The pinch strength was measured using the SAEHAN pinch gauge. Three different measurements were taken using the pinch gauge: key pinch, tip pinch and palm pinch, as shown in Fig. 2. The pinch strength measurements were also taken before and after each bike riding trial. The hand strength variation was recorded as the difference between pinch measures before and after bike riding.

2.2 Analysis

For analyzing data Statistical Packages for the Social Science (SPSS) software was used. The data collected from the hand discomfort survey were analyzed using descriptive statistics. Correlation between hand discomfort and average riding hours per day, BMI and total bike riding experience of the person was determined using Spearman's correlation. Two-way ANOVA with repeated measures was used to analyze the experimental data, which aims to identify the interaction between subject factor and dependent variable. It was also used to compare the mean differences between different demographic groups of subjects.

3 Results and Discussions

3.1 Discomfort Survey

The prevalence of hand discomfort among bike riders was confirmed by the result of the discomfort survey. The percentage of participants who felt discomfort in various regions of the hand is shown in Fig. 3. The majority of participants (45%) felt discomfort in the ring, middle and index fingers of the left hand. 20% of participants responded that they felt discomfort on the hypothenar muscle of the right hand. 17% of participants responded that they felt discomfort on the palm of both hands. A comparatively lesser percentage of participants reported discomfort in other regions. In the figure, the hand regions A to F are suffixed as second letter with the left (L) or right (R) hand denotations as the first letter in abbreviations.

The frequency of discomfort among subjects in different regions of the hand is shown in Fig. 4. Among subjects who felt discomfort on the ring, middle and index fingers of the left hand, 26 of them experienced discomfort 1–2 times in that week and 19 of them experienced discomfort 3–4 times in the same time span. Among subjects who felt discomfort on the hypothenar muscle in the right hand, 13 subjects experienced discomfort 1–2 times in that week and 7 experienced discomfort 3–4 times in that week. Among participants who felt discomfort on the palm of right hand, 14 of them experienced discomfort 1–2 times in that week and 5 experienced

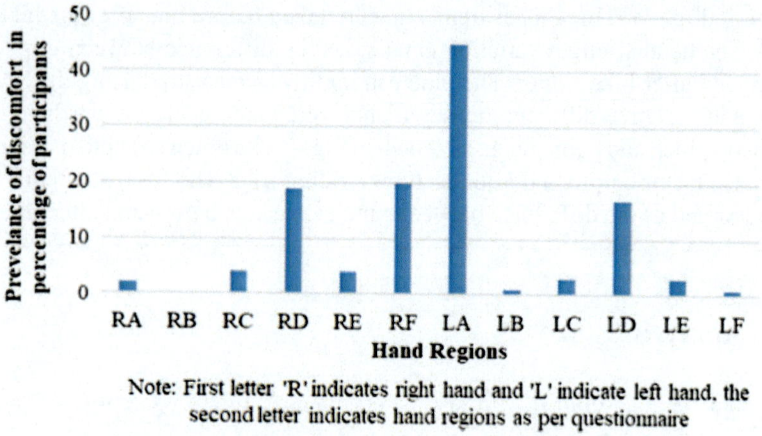

Fig. 3 Prevalence of discomfort among bike riders

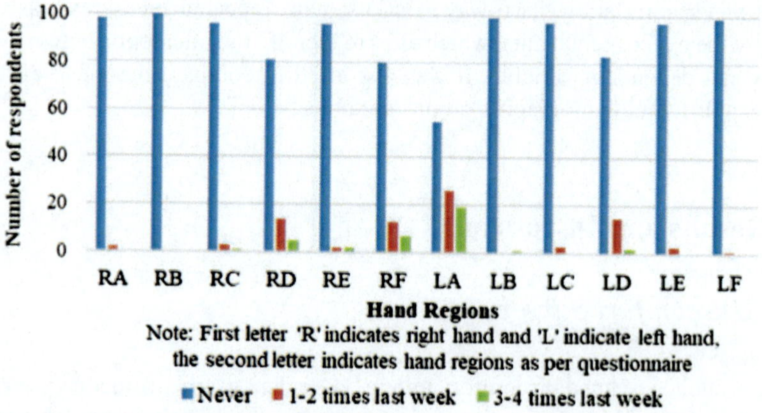

Fig. 4 Frequency of discomfort among bike riders

discomfort 3–4 times in that week. A total of 15 participants responded that they experienced discomfort 1–2 times in that week on the palm of left hand and only 2 have experienced discomfort 3–4 times in that week. From the result it is evident that the discomfort is not frequent among the bike riders. The discomforts do not occur in a daily pattern and the peak value of frequency of discomfort is 3–4 times per week as per this study.

In the discomfort survey, no participants responded that the discomfort interfere with their ability to work. Only three participants responded that the discomfort is moderately uncomfortable and none responded that discomfort is very uncomfortable. The discomfort was slightly uncomfortable in all other's opinions. The discomfort score was calculated and the total discomfort score of all regions was determined. The mean total discomfort score in the age groups 20–24, 25–29, 30–34 and 35–39 years are 1.90, 2.68, 3.00 and 3.06, respectively. The mean score of the

Table 1 Spearman's correlation of parameters with total discomfort score

Parameters	Spearman's correlation with total discomfort score
BMI	−0.200*
Riding experience	0.067
Average riding hour per day	0.453*

* $p < 0.05$

total hand discomfort was observed to be less in the age group of 20–24 years and high in the age group of 35–39 years. The discomfort is observed to be increasing with age. The maximum discomfort score observed among the participants is also from the higher age groups.

The Spearman's correlations were determined between total discomfort score and participants BMI, total bike riding experience and average riding hours per day. Statistically significant ($p < 0.05$) weak negative correlation of—0.200 was obtained between BMI and the total discomfort score (Table 1). As per this result, the hand discomfort will be less for persons with higher BMI. The correlation between bike riding experience and discomfort score was insignificant ($p > 0.05$). A positive correlation of 0.453 between average riding hours a day and discomfort score was found significant ($p < 0.05$). As per this result, increased riding hours a day could cause more hand discomfort.

3.2 Field Experiment

Marginal means of variation in grip strength and pinch strengths are shown in Table 2. The variation was calculated as the difference between hand strength before riding and hand strength after riding.

Grip strength shows a negative value in mean variation which means that the grip strength value obtained after riding was greater than the initial value. The variation

Table 2 Marginal means of grip strength, key pinch strength and palmar pinch strength variation

Road condition	Use of gloves	Mean hand strength variation in newton					
		Grip strength		Key pinch strength		Palmar pinch strength	
		Left hand	Right hand	Left hand	Right hand	Left hand	Right hand
Good	No	−26.151	−44.044	6.783	13.764	3.351	8.826
	Yes	−16.998	−18.237	3.432	6.882	0.49	2.615
Bad	No	−14.383	−32.001	10.46	19.097	6.047	12.994
	Yes	−19.613	−22.022	3.759	9.721	−0.082	5.557

Table 3 Effect of road conditions and gripping conditions in hand strength variation

	ANOVA test statistics (F values)					
	Grip strength		Palmar pinch strength		Key pinch strength	
Source	Left hand	Right hand	Left hand	Right hand	Left hand	Right hand
Road conditions	1.27	1.56	0.59	3.74	4.91*	24.57*
Gloves	0.16	16.67*	10.90*	20.31*	38.17*	106.10*
Road conditions × Gloves	4.31	3.05	2.47	0.69	8.11*	4.78*

* $p < 0.05$

can be explained as muscle activation and stretching during bike riding. Literature indicates that muscle activation and stretching cause increased muscle strength (Feldman 1986). The positive value of variation in pinch strength indicates a decrease in hand strength during bike riding which is a direct indication of muscle fatigue. Tip pinch strength measurements could not provide any significant results.

ANOVA results of the effect of road conditions and gripping conditions on hand strength variation during bike riding are shown in Table 3. From the results, both road condition and gripping condition were found to have a statistically significant effect on right-hand and left-hand key pinch strength variation. The key pinch strength variations were minimal when riding with gloves in good road condition. In bad road condition increased vibration may cause muscle fatigue. In this study, fatigue is maximal when riding with bare hands in bad road condition. Also, the results indicate that the use of hand gloves helps to decrease muscle fatigue. The effect may be because the gloves act as a vibration damper.

In the case of grip strength measurements, riding with hand gloves has shown a significant effect ($p < 0.05$) on grip strength variation of the right hand. It was also found that hand strength is increasing during bike riding. It may have occurred due to the impact of static muscle force. Static muscle force depends on muscle length due to stretch reflex and elasticity of active muscle fibers. If the static muscle force only affects the muscle length, then it would behave like an ordinary spring changing its equilibrium length that is only dependent on the magnitude of the load (Feldman 1986). The use of hand gloves reduces the increase in hand strength or muscle stretching and activation.

From the ANOVA results, only hand gloves were found to have a statistically significant effect on palmar pinch strength variation. The palmer pinch strength was decreased during bike riding and it was found that the variation can be reduced by using hand gloves. Tip pinch strength variation of both hands was found insignificant ($p > 0.05$) in all the cases.

4 Conclusion

The survey and experiment were carried out to determine hand discomforts and causes of hand strength variations, which arise due to muscle fatigue or muscle stretching, or both. From the survey, it was found that a higher percentage of people experience hand discomforts due to bike riding. The most affected regions of hand for different age groups are identified and the results were almost the same in all age groups. The problems were the same but the severity was different. No one responded that problem affects their ability to work or discomfort is very uncomfortable. The frequency of discomfort was never more than 3–4 times a week, hence it can be concluded that problems are not severe.

The field experiment was aimed to study the strength variation during bike riding. Two independent factors, road condition and hand gripping condition, were selected. The results of the experiment revealed that there is an increase in strength in the case of grip strength measurements and a decrease in strength in the case of key pinch and palmar pinch. The decrease in strength indicates muscle fatigue due to bike riding. The increase in strength indicates increased muscle activation and muscle length. The use of gloves finds useful in limiting the variation of strength. The road condition is a cause of variation in key pinch strength. In bad road condition more key pinch strength variations occurred. It reveals that the bad road condition is causing more muscle fatigue. For comfort riding good road condition is inevitable.

The problems identified and correlations obtained from the study can be used to design countermeasures of hand discomfort due to bike riding. The study can be extended to a larger population by including old age people. Nowadays, they are also using two-wheelers for daily transport. There is only limited research work carried out in the field of hand strength variation. Hence, more researches are required to enrich this field. It could be useful in the development and design of hand tools and other hand-operated devices.

References

Berolo S, Wells RP, Amick BC (2011) Musculoskeletal symptoms among mobile hand-held device users and their relationship to device use: a preliminary study in a Canadian university population. Appl Ergon 42(2):371–378

CUergo: Hand Discomfort Questionnaires. http://ergo.human.cornell.edu/ahhandmsquest.html. Accessed 02 Mar 2018

Dagli R (2016) Driver fatigue and road safety—implication in an indian context. Int J Adv Eng Technol 9(4):502–506

Feldman AG (1986) Once more on the equilibrium-point hypothesis (λ model) for motor control. J Mot Behav 18(1):17–54

Hedge A, Morimoto S, Mccrobie D (1999) Effects of keyboard tray geometry on upper body posture and comfort. Ergonomics 42(10):1333–1349

Kong YK, Song YW, Jung MC Lee I (2011) Effects of hand position on maximum grip strength and discomfort. In: HFESA 47th annual conference. Curran Associates Inc., New York, pp 1–5

Mathiowetz V, Weber K, Volland G, Kashman N (1984) Reliability and validity of grip and pinch strength evaluations. J Hand Surg 9(2):222–226

Muscles of the hand. https://www.revolvy.com/page/Muscles-of-the-hand. Accessed: 23 Jun 2018

Open Government Data (OGD) Platform India. Total Number of Registered Motor Vehicles in India. https://data.gov.in/catalog/total-number-registered-motor-vehicles-india. Accessed 09 Apr 2018

Peters MJ, van Nes SI, Vanhoutte EK, Bakkers M, van Doorn PA, Merkies IS, Faber CG (2011) PeriNomS study group: revised normative values for grip strength with the Jamar dynamometer. J Peripher Nerv Syst 16(1):47–50

Pibphoto.nic.in. http://pibphoto.nic.in/documents/rlink/2016/jun/p20166905.pdf. Accessed 09 Apr 2018

Reis MM, Arantes PMM (2011) Assessment of hand grip strength—validity and reliability of the saehan dynamometer. Fisioterapia e Pesquisa 18(2):176–181

Robertson LD, Mullinax CM, Brodowicz GR, Swafford AR (1996) Muscular fatigue patterning in power grip assessment. J Occup Rehabil 6(1):71–85

Schwid SR, Thornton CA, Pandya S, Manzur KL, Sanjak M, Petrie MD, McDermott MP, Goodman AD (1999) Quantitative assessment of motor fatigue and strength in MS. Neurology 53:743–750

Swanson AB, Matev IB, de Groot G (1970) The strength of the hand. Bull Prosthet Res 10(14):145–153

Vollestad NK (1997) Measurement of human muscle fatigue. J Neurosci Method 74(2):219–227

Chapter 12
A Study on Waste Loading Task and Influence of Waste Container Size

Francis J. Emmatty and Vinay V. Panicker

1 Introduction

Waste collection workers have a vital role in the recycling process and facilitate a sustainable environment. Workers in the waste collection are vulnerable to the highest risk of injuries and illness compared with other occupations (UN-Habitat 2010; Çakit 2015). The lack of sufficient facilities and equipment puts a tremendous burden on waste collection workers (Yang et al. 2001). Compared to the workers in developed countries, workers in third world countries are subject to increased occupational risks. These workers are not adequately paid even in developed nations.

Waste collection workers frequently do physical tasks such as lifting, carrying, pulling and pushing (Yang et al. 2001; Jeong 2016). Occupations involving heavy, repetitive tasks and awkward postures are associated with musculoskeletal disorders (MSDs) (Engkvist et al. 2011; Bleck and Wettberg 2012; Singh and Chokhandre 2015; Zakaria et al. 2017). Waste collection activities may result in damage to the body parts, including the back, knees, hands and feet (Ivens et al. 1998). The most commonly occurring injury is in the low back region (Poulsen et al. 1995).

In developing countries, waste collection workers perform lifting and tasks manually (Gutberlet et al. 2013; Mol et al. 2017). The twisting and bending of the body during lifting tasks in waste-related occupations result in musculoskeletal illnesses (Jeong et al. 2016). In developed countries, automatic lifting devices are used to empty waste containers (Kuijer and Frings-Dresen 2004), and there is a need for such lifting devices in developing countries for waste collection (Jeong et al. 2016).

The physical workload and heart rate are higher for waste loading workers compared to other workers involved in the waste collection (Tsujimura et al. 2012). A laboratory study in the USA evaluated the physical load of waste collection tasks

F. J. Emmatty · V. V. Panicker (✉)
Industrial Engineering Laboratory, Department of Mechanical Engineering, National Institute of Technology Calicut, Calicut, Kerala, India
e-mail: vinay@nitc.ac.in

© The Author(s), under exclusive license to Springer Nature Singapore Pte Ltd. 2022, corrected publication 2022

D. Bijulal et al. (eds.), *Technology-Enabled Work-System Design*, Design Science and Innovation, https://doi.org/10.1007/978-981-16-1884-0_12

and stated that lifting and dumping are very high-risk activities (Çakit 2015). In the above research, six subjects simulated loading tasks with a safe load of 9.5 kg, and the tasks are evaluated using Rapid Upper Limb Assessment (RULA), Rapid Entire Body Assessment (REBA) and Digital Human Modelling (DHM). A standardized Nordic questionnaire-based study in India reported that waste loaders are prone to the risk of MSDs and related disabilities (Salve et al. 2017). A posture assessment study from India found that the postures of waste collection workers have to be changed immediately (Emmatty and Panicker 2019). These workers are vulnerable to physical injuries, such as back pain and hand and leg joint discomfort.

The occupational risks of these workers can be mitigated by redesigning the containers for manual loading. The container properties to be reconfigured in the design process for manual handling are weight, distribution of weight, geometry, stiffness and availability of coupling devices like handles (Ayoub 1977). No studies have been found in the literature reporting study of waste loading tasks by DHM, among workers in India. The influence of bin geometry on the lifting and dumping tasks was also not found in the literature.

This study aims to assess occupational risks among Indian waste loaders. The postures of waste loaders are evaluated by RULA and are compared with RULA scores of digital worker models generated in DHM. The study further assesses the influence of waste bin geometry on the physical workload of the workers by simulating the compressive forces at L4/L5 intervertebral disc. The impact of factors including gender, percentile height and percentile weight on the compressive force is also discussed in this study.

After the introduction and literature review, in the second section, the methodology for solving the problem description is given. In the third section, the results of posture evaluation and variation of compressive forces on L4/L5 vertebrae are analysed. In the fourth and final section, the discussion on the results and conclusion is provided.

2 Methodology

Work-study of waste loading workers in an educational institution is performed by video motion analysis. By using two cameras, the footage is taken to track the two male workers involved in the waste loading activities. The activity of waste loading is classified into ten work elements. RULA posture scores for each of the ten work elements are found by extraction of frames of video considering the most awkward posture for each of the work elements. RULA is a postural screening tool for evaluating the vulnerability of workers to risk factors related to upper extremity activities, classifying body postures and recognizing musculoskeletal risks (McAtamney and Corlett 1993).

Digital human models are simulated similar to that of real workers using Siemens Jack DHM software (Siemens 2015). Digital worker models are generated from the Indian anthropometric data (Chakrabarti 1997) using DHM. The virtual environment is created considering the actual work environment of the workers including the height of the vehicle, dimensions of the waste bin and the position of the workers doing the task.

The digital worker models are validated by comparing the postures of waste loading workers in the video with that of digital worker models generated. A comparison of the RULA posture scores of the workers in the video and that of the digital worker models ensure that the postures are modelled to some degree of accuracy in the DHM. The RULA analysis using direct observation and DHM has identified lifting as a critical work element.

The compressive force at L4/L5 intervertebral disc of the lumbar spine during lifting is simulated using DHM, and the influence of bin geometry (normal/small size) on compressive force is studied. The dimensions of the bin are scaled down to 0.6 times along the three axes to get a small bin. It is assumed that the change in the geometry of the bin will allow the worker to select a convenient posture to do the lifting task. The average weight of a bin filled with waste is approximately 20 kg for both normal and small size bin, and an asymmetric loading with 25% of the weight of the bin is assumed to act on the left palm centre and 75% of the weight of the bin is on the right palm centre of the worker model.

Compressive forces at L4/L5 intervertebral disc are simulated by changing the following four factors: bin geometry (normal/small size), gender (male/female), percentile height (5th, 50th and 95th) and percentile weight (5th, 50th and 95th); i.e. 36 combinations and for each of these combinations five iterations have been carried out in Force solver tool in Jack simulation software. A statistical analysis using ANOVA has been carried out to test the influence of each of these four factors on the compressive forces at L4/L5 intervertebral disc.

3 Results and Discussion

3.1 Posture Evaluation

The posture evaluation is conducted for the ten work elements for the waste loading task using RULA by direct observation and using DHM. RULA posture score of waste loading task by observation and DHM is given in Table 1. A male worker (M1) transports the bin filled with waste and lifts the container to another worker (M2). Worker M2 collects the bin standing within the vehicle, unloads the waste into the vehicle and transfers the bin to worker M1. The photograph of the worker performing the lifting task of the normal bin is given in Fig. 1. The illustration of the digital worker model (male) performing the lifting task of the normal bin is provided in Fig. 2.

In the case of waste loading task, the posture score obtained by observation of video and the posture score generated by DHM are almost the same for most of the activities. The scores for non-critical tasks differ slightly in some of the work elements as these activities can be carried out differently by the workers. Thus the

Table 1 RULA posture score of waste loading task by observation and DHM

Activity description	RULA score—M1		RULA score—M2	
	By observation	By DHM	By observation	By DHM
1. M1 transports the bin up to the vehicle (M2 idle)	2	2	1 (idle)	1 (idle)
2. M1 aligns the bin (M2 idle)	4	4	1 (idle)	1 (idle)
3. M1 opens the lid of the bin (M2 idle)	3	3	1 (idle)	1 (idle)
4. M1 lifts the bin (M2 idle)	7	7	1 (idle)	1 (idle)
5. M1 tilts the bin and M2 collects the bin	6	6	6	6
6. M2 dumps and shakes the bin (M1 idle)	1 (idle)	1 (idle)	7	7
7. M2 transfers the bin and M1 receives the bin	4	4	4	4
8. M1 lowers the bin (M2 idle)	5	5	1 (idle)	1 (idle)
9. M1 closes the lid of the bin (M2 idle)	4	3	1 (idle)	1 (idle)
10. M1 transports the bin back to the stand	3	4	1 (idle)	1 (idle)

Fig. 1 Photograph of the worker performing the lifting task of the normal bin

digital worker models are validated based on the similarity of the postures. Lifting and dumping activities are identified as critical work element with a RULA score of 7, indicating very high risk level for the workers (shown as shaded in Table 1).

Fig. 2 Digital worker model (male) performing the lifting task of the normal bin

3.2 Compressive Forces at L4/L5 Intervertebral Disc

The lifting of the bin of 20 kg is simulated for determining the forces at L4/L5 intervertebral disc for both normal and small bin. The posture for lifting a normal bin is as shown in Fig. 2 and DHM takes a posture for lifting a small bin as shown in Fig. 3. The plot of compressive forces at L4/L5 for different bin geometry and gender is given in Fig. 4. The compressive forces while lifting the small bin is less compared to the normal bin, and the means of compressive forces are statistically different ($\alpha = 0.05$). The reduction in compressive forces is due to the change from a posture requiring flexing of the back to a squatting posture.

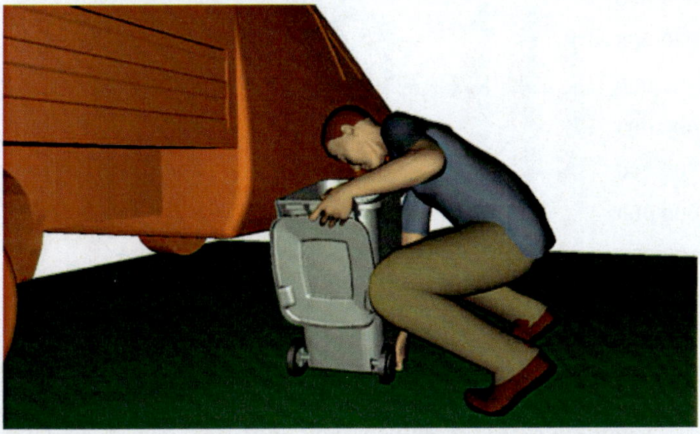

Fig. 3 Digital worker model (male) performing the lifting task of the small bin

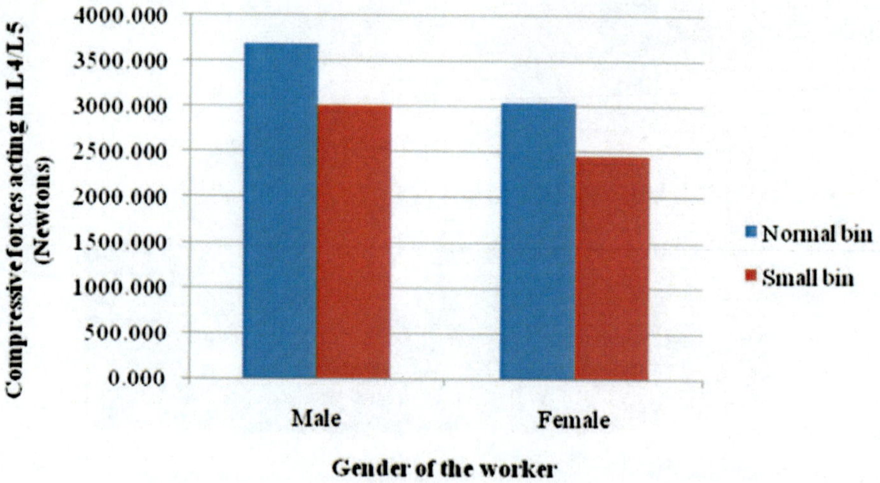

Fig. 4 Plot of compressive forces at L4/L5 for different bin geometry and gender

The plot of compressive forces at L4/L5 for male digital workers for different percentile weight and percentile height is given in Figs. 5 and 6. The observation of Fig. 5 reveals that the reduction in compressive forces while lifting a small bin is more for male workers of 95th percentile height compared to that of 5th percentile height since 95th percentile male workers are affected more by flexing of the back while lifting normal bin. It is observed that compressive forces at L4/L5 are above

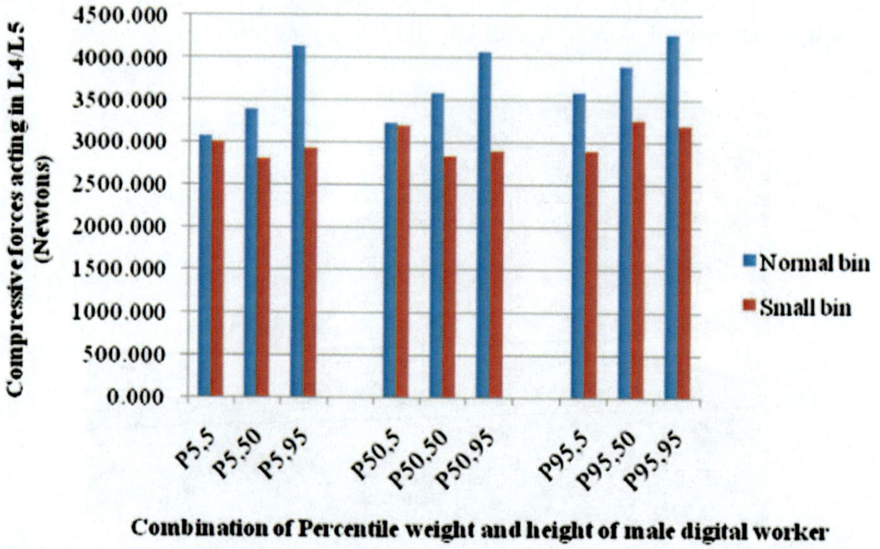

Fig. 5 Plot of compressive forces at L4/L5 for male digital workers for different percentile weight and height (Pi,j = Combination of 'i'th percentile weight and 'j'th percentile height)

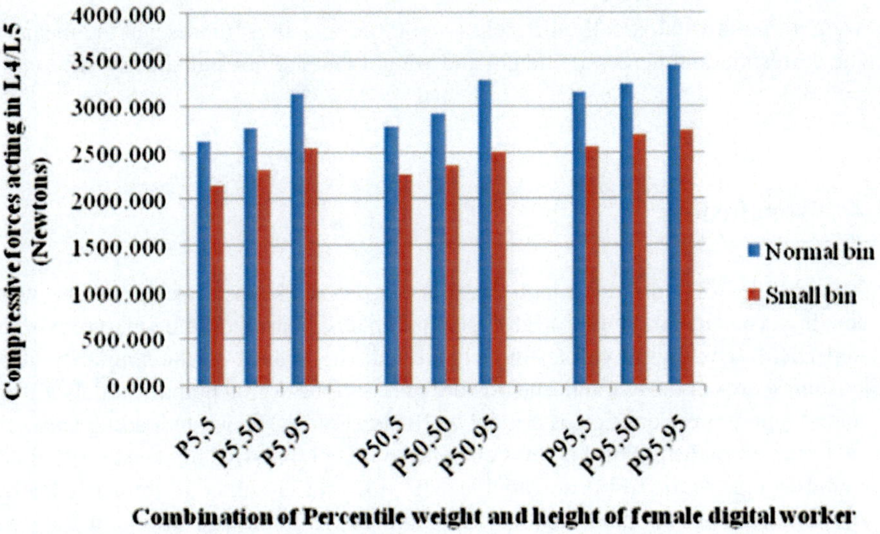

Fig. 6 Plot of compressive forces at L4/L5 for male digital workers for different percentile weight and height (Pi,j = Combination of 'i'th percentile weight and 'j'th percentile height)

the NIOSH recommended limit of 3400 N (Waters et al. 1993) and the change in the bin geometry has reduced the compressive forces to an acceptable range. It is also observed that gender, percentile height and percentile weight are statistically significant factors ($\alpha = 0.05$). As the factors of percentile height and percentile weight are significant, a pairwise comparison of means is done by Tukey's HSD test for the lifting of the normal/small bin by male/female workers as given in Table 2. In Table 2, 'S' represents the means of the corresponding percentile height and

Table 2 Pairwise comparison of means by percentile height and percentile weight

	Normal bin					
	Percentile height			Percentile weight		
	5th = 50th	50th = 95th	5th = 95th	5th = 50th	50th = 95th	5th = 95th
Male	S	S	S	NS	NS	S
Female	NS	S	S	NS	S	S
	Small bin					
	Percentile height			Percentile weight		
	5th = 50th	50th = 95th	5th = 95th	5th = 50th	50th = 95th	5th = 95th
Male	NS	NS	NS	NS	NS	S
Female	NS	NS	S	NS	S	S

'S' = the means of the corresponding percentile height and weight are statistically different ($\alpha = 0.05$) and 'NS' = the means of the corresponding percentile height and weight are not statistically different ($\alpha = 0.05$).

weight that are statistically different ($\alpha = 0.05$) and 'NS' represents the means of the corresponding percentile height and weight that are not statistically different ($\alpha = 0.05$).

4 Conclusion

The chances of musculoskeletal hazards for the work postures during the waste loading tasks are studied by RULA posture analysis based on the video of workers and DHM. The digital worker models are validated based on the similarity of the posture scores of the workers and posture scores of the digital human models. Lifting and dumping are identified as critical work elements in the waste loading task.

In this study lifting of the bin is analysed with compressive forces at L4/L5 intervertebral disc as the response variable. The study has assumed asymmetric loading on both hands at the palm centre. The compressive forces are lower for the small bin compared to the normal bin both of the same weight. The research reveals that by improving the geometry of the bin, the worker can select better postures, thereby reducing the compressive forces at L4/L5.

Further research is needed to determine the optimum design parameters for efficient handling of the bin to improve the posture of workers involved in waste loading tasks. This analysis has only considered the influence of bin geometry on physical workload. There is also scope for assessing the physical workload of the dumping activity. The study did not measure the distribution of the forces acting on both hands of the worker to provide realistic inputs to the DHM.

Acknowledgements The first author acknowledges the Directorate of Technical Education Kerala, India and All India Council for Technical Education, India for financial support for research under the Quality Improvement Programme (QIP) scheme.

References

Ayoub MA (1977) Optimum design of containers for manual material handling tasks. Appl Ergon 8(2):67–72

Bleck D, Wettberg W (2012) Waste collection in developing countries—tackling occupational safety and health hazards at their source. Waste Manag 32(11):2009–2017

Çakit E (2015) Assessment of the physical demands of waste collection tasks. Glob Nest J 17(2):426–438

Chakrabarti D (1997) Indian anthropometric dimensions for ergonomic design practice. National institute of design

Emmatty FJ, Panicker VV (2019) Posture assessment among waste loading workers in India. In: Vasudevan H, Kottur V, Raina A (eds) Proceedings of international conference on intelligent manufacturing and automation. Lecture notes in mechanical engineering. Springer, Singapore

Engkvist I-L, Svensson R, Eklund J (2011) Reported occupational injuries at Swedish recycling centres—based on official statistics. Ergonomics 54(4):357–366

Gutberlet J, Baeder AM, Pontuschka NN, Felipone SMN, dos Santos TLF (2013) Participatory research revealing the work and occupational health hazards of cooperative recyclers in Brazil. Int J Environ Res Public Health 10(10):4607–4627

Ivens UI, Lassen JH, Kaltoft BS, Skov T (1998) Injuries among domestic waste collectors. Am J Ind Med 33(2):182–189

Jeong BY (2016) Occupational injuries and deaths in domestic waste collecting process. Hum Fact Ergon Manuf 26(5):608–614

Jeong BY, Lee S, Lee JD (2016) Workplace accidents and work-related illnesses of household waste collectors. Saf Health Work 7(2):138–142

Kuijer PPFM, Frings-Dresen MHW (2004) World at work: refuse collectors. Occup Environ Med 61(3):282–286

McAtamney L, Corlett EN (1993) RULA: a survey method for the investigation of work-related upper limb disorders. Appl Ergon 24(2):91–99

Mol MPG, Pereira AF, Greco DB, Cairncross S, Heller L (2017) Assessment of work-related accidents associated with waste handling in Belo Horizonte (Brazil). Waste Manag Res 35(10):1084–1092

Poulsen OM, Breuma NO, Ebbehøj N, Hansena M, Ivensa UI, Van Lelieveld D, Matthiasen' L, Nielsena BH, Nielsena EM, Schibye B, Skov T, Stenbaeka EI, Wilkins CK (1995) Sorting and recycling of domestic waste. Review of occupational health problems and their possible causes. Sci Total Environ 168(1)33–56

Salve P, Chokhandre P, Bansod D (2017) Assessing musculoskeletal disorders among municipal waste loaders of Mumbai, India. Int J Occup Med Environ Health 30(6):875–886

Siemens (2015) Jack user manual version 8.3. Siemens product lifecycle management software Inc. Hope Street Cypress, California USA

Singh S, Chokhandre P (2015) Assessing the impact of waste picking on musculoskeletal disorders among waste pickers in Mumbai, India: a cross-sectional study. BMJ Open 5(9):e008474

Tsujimura H, Taoda K, Kitahara T (2012) A field study on the physiological workload of garbage collectors in the Japanese summer. Ind Health 50(6):556–566

UN-Habitat (2010) Solid waste management in the world cities. United Nations human settlements programme

Waters TR, Putz-Anderson V, Garg A, Fine LJ (1993) Revised NIOSH equation for the design and evaluation of manual lifting tasks. Ergonomics 36(7):749–776

Yang C-Y, Chang W-T, Chuang H-Y, Tsai S-S, Wu T-N, Sung F-C (2001) Adverse health effects among household waste collectors in Taiwan. Environ Res 85(3):195–199

Zakaria J, Eh S, Fac O, Nfm S (2017) Musculoskeletal disorder among municipal solid waste collectors. Asia Pacific Environ Occup Heal J 3(1):28–32

Chapter 13
Postural Assessment of a Load-Lifting Task for Females: A Comparative Assessment of OWAS and REBA Method

Sumaiya, M. Shah Faizan, and Mohammad Muzammil

1 Introduction

Although automation has paved the way in industries for material handling, it cannot replace humans as they are faster, flexible, and have a low cost of hiring. Manual material handling (MMH) is one of the most common activities in the workplace causing over one-third of all workplace injuries (Health and safety statistics for manufacturing sectors in Great Britain 2017). Load-lifting tasks mainly affect the neck, shoulder, hands, and lower back. Occupational epidemiology shows that manual material-handling tasks are the major cause of lower back pain and several other related musculoskeletal disorders (Zurada 2012). The injuries caused by MMH are mainly due to awkward posture, the frequency of lift, heavy load, repetitive movements of arms, etc. Awkward working posture is one of the leading causes of musculoskeletal disorders which may be short-lived such as strain, sprain, or maybe a cause of lifelong pain and disability.

To reduce the effect of musculoskeletal problems associated with the load-lifting task, it is vital to understand the effect of task parameters on human posture. The different methods to measure working postures are photographic techniques, inclinometers, goniometers, electro-goniometers, and video recording systems. Apart from these postural assessment techniques, electromyography analysis of the affected muscle may also be an option to quantize the risk of load-lifting tasks in terms of fatigue which may be used to design different material-handling tasks to prevent any kind of musculoskeletal disorder (Mo and Jung 2015). A dynamic 3D-linked segment model with a detailed electromyography-driven model of truck musculature was used to study the effect of lifting height and load mass on low back loading concluding that with an increase in lifting height, the maximum low back moment decreased (Hoozeman et al. 2008). The literature survey shows that most of the investigations of workers and workplaces use subjective measurement techniques

Sumaiya (✉) · M. S. Faizan · M. Muzammil
Mechanical Engineering Department, Aligarh Muslim University, Aligarh, India

© The Author(s), under exclusive license to Springer Nature Singapore Pte Ltd. 2022, corrected publication 2022

D. Bijulal et al. (eds.), *Technology-Enabled Work-System Design*, Design Science and Innovation, https://doi.org/10.1007/978-981-16-1884-0_13

to correlate different physical factors. Subjective measurement techniques are not reliable because the reading varies according to the perception of the subjects (Liu 2014). In this study two quantitative postural assessment techniques (OWAS and REBA) have been used to overcome the disadvantages of subjective methods.

Recent studies show remarkable growth in the female participation of the labor force globally (Female labor supply 2018) but studies associated with female workers are very less in number. Karwowski (1991) has evaluated the subjective perception of load heaviness by females and related this perception to the maximum acceptable weights of limits (Karwowski 1991). In a comparative study of maximum weight lift by a male and female industrial worker by (Mital 1984, 2007), it is shown that men significantly lift more weight than females for a 12-h shift, and the metabolic energy expenditure rate of females is also less (Mital 2007). The effect of load and posture on load estimations during a simulated lifting task is studied among female check-out operators, which shows the independent and interactive effect of load and postures during a simple load-handling task (Butler and Kosey 2003). The above studies have led to a focus on female industrial workers involved in MMH.

In the present study, an effort has been made to carry out postural analysis for female workers performing load-lifting tasks. The Ovako Working Posture Assessment (OWAS) is carried out with the help of the Kinect motion sensor integrated with Ergonautas software while Rapid Entire Body Assessment (REBA) is done using an employee worksheet (Hignett and McAtamney 2000).

2 Method

2.1 Participants

Ten young healthy female subjects having age 22 ± 0.83 years, weight 60 ± 2.06 kg, and height 159 ± 1.65 cm volunteered for the experiment. The recruitment criteria for the subjects included the absence of any kind of musculoskeletal disorder, lower back pain, general illness, balance problems, and apparent postural deviations.

2.2 Procedure

The experimental setup for the load-lifting task is shown in Fig. 1. The task included the lifting and shifting of boxes weighing 4, 6, and 8 kg from a table height of 30 cm (on the left side of the subject) to the three different table heights of 60, 80, and 100 cm (on the right side). The position of the foot of the subject is kept static in the center of the two tables which is 60 cm apart. Since the task involves the shifting of boxes from the left side of the body to the right, so it is accompanied by both bending and twisting of the trunk. The selection of the various parameters was based

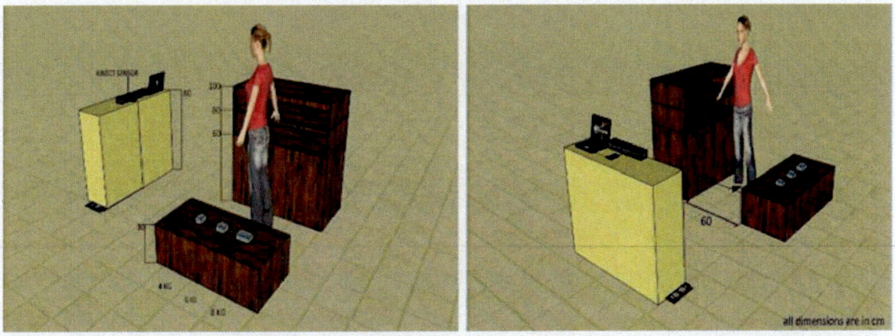

Fig. 1 Experimental setup for a load-lifting task

on Karwowski's (1991) study of psychophysical acceptability and perception of load heaviness by the subjects (Karwowski 1991). Twenty-seven sets of reading for each subject were taken, repeating each lifting task thrice. Accounting for the sensitivity of the Kinect sensor, subjects were briefed to maintain a constant speed during the lifting task and to maintain continuous eye contact with the camera.

2.3 Data Collection

The OWAS analysis is done with the help of Ergonautas-NUI software integrated with the Kinect motion sensor. Kinect sensor is a cost-effective 3D motion sensor that is more sensible in achieving a natural interaction with the user. It can retrieve depth data (third dimension), color data (RGB data), and information about the position of the joints of the recognized user present in the frame (skeletal data). The information of skeletal data is captured in the form of an array with 20 coordinates of joints. The joint centers are captured following a randomized decision forest algorithm (Mas and Marzal 2014) and it seems to be accurate to be used for postural assessment (Dutta 2012). At a regular interval of time, Ergonautas-NUI together with the OWAS application records the skeletal tracking joint information and the data collected is processed to obtain the codes for each body postures and risk action level as shown in Fig. 2.

For REBA, 11 snapshots have been taken from the video recorded during the load-lifting task as shown in Fig. 3. REBA employee assessment worksheet has been used for the analysis part (Hignett and McAtamney 2000).

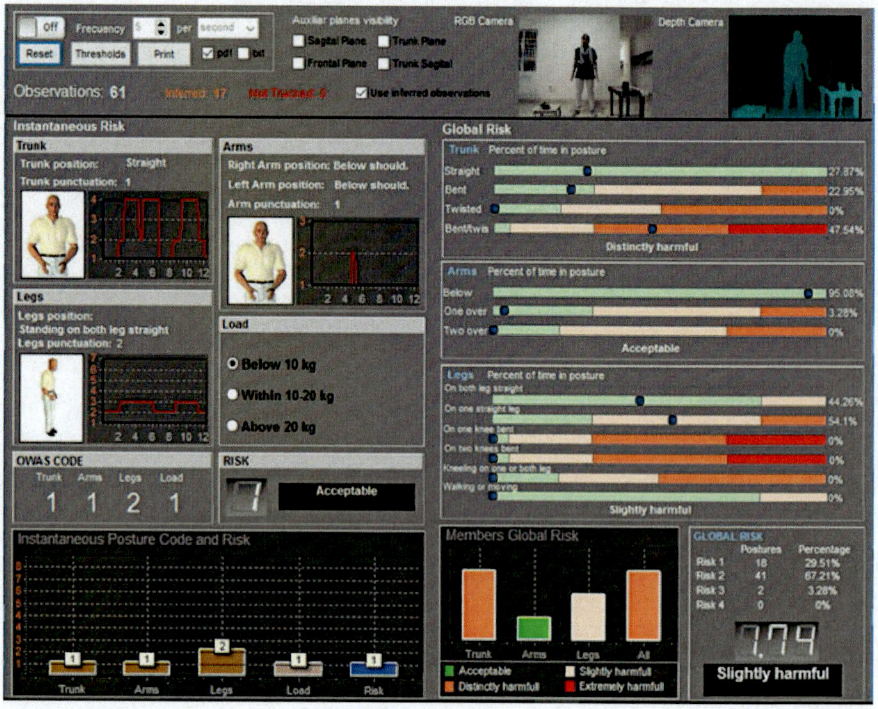

Fig. 2 Sample result of OWAS assessment for a load of 8 kg and table height of 60 cm

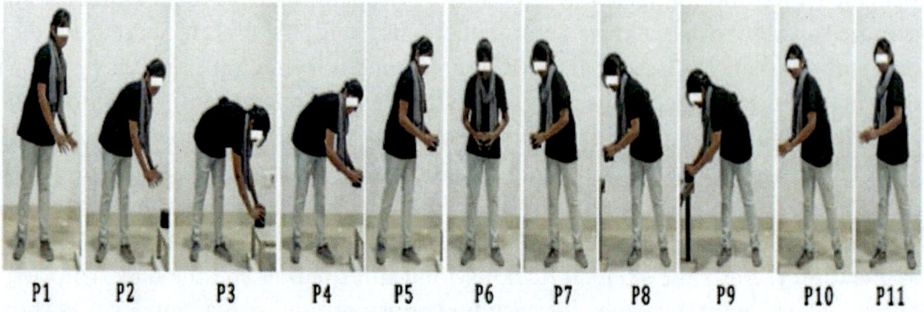

Fig. 3 Different positions acquired during the load-lifting task

2.4 *OWAS Analysis*

The overall movement of a subject is captured by the Kinect sensor which simultaneously transfers data to Ergonautas software. The Ergonautas with the help of OWAS methodology gives the instantaneous reading of the risk associated with the posture and the overall global risk rating of the task performed. Figure 2 shows the sample output given by the Ergonautas software while lifting a load of 8 kg to the table height of 60 cm. The frequency of capturing is set at 5 frames per second. It gives readings in two parts instantaneous (left) and the global rating (right). The member's global risk rating is given separately for trunk, arms, and legs.

2.5 REBA Analysis

A score is assigned to various body regions like the neck, trunk, legs, and arms (lower and upper) mentioned in the REBA sheet. The scoring process also involves the inclusion of several other factors like twisting, bending, abduction, force, coupling, activity score, etc. After the data for each region is collected and scored, tables on the form are then used to compile the risk factor variables, generating a single score that represents the level of MSD risk. The scoring is as follows:

Score 1	Negligible risk.
Score 2, 3	Low risk, changes may be needed.
Score 4–7	Medium risk, further investigation, and change soon.
Score 8–10	High risk, investigate, and implement change.
Score 11+	Very high risk, implement change.

3 Results

The results of the OWAS assessment done using the Kinect sensor are shown in Table 1. Kinect sensor captured approximately 60 postures of a subject during each lifting task. The Ergonautas software integrated with the Kinect sensor classifies these postures following OWAS methodology into different risk categories. A global risk rating is generated considering the risk associated with each posture involved in the task. The results show that the lifting task performed for the varying load and height conditions is under the risk category of slightly harmful. With the increase in load, the global risk associated with the task increases. For a particular load, the global risk rating is minimum for 80 cm and maximum for 100 cm table height. Further, the member's global risk for all the varying conditions is the same, with the trunk under the risk category of distinctly harmful, legs under the risk category of slightly harmful and arms under the acceptable range.

The results of the postural assessment of the load-lifting task evaluated using the REBA method for different loads and positions are shown in Figs. 4, 5, and 6 (also shown in Fig. 3). It is evident from the graph that the REBA score of postures P3, P8, and P9 is above 7, thus, it is associated with high risk, while posture P4 is under the medium risk (based on REBA sheet guidelines). Moreover, it is also evident from the graph that the variation of load and the table height has very little effect on the REBA score. However, the REBA score is slightly high for some posture that involves load placing at the 60 cm table height.

Table 1 The risk associated with posture for different table height and load conditions

Load lifted (kg)	Table height (mm)	Posture frequency				Global risk rating	
		Risk 1	Risk 2	Risk 3	Risk 4		Risk degree
4	60	15	36	0	0	1.71	Slightly harmful
4	80	35	41	0	0	1.54	Slightly harmful
4	100	31	42	0	0	1.58	Slightly harmful
6	60	18	45	0	0	1.73	Slightly harmful
6	80	26	36	0	0	1.62	Slightly harmful
6	100	16	30	0	0	1.65	Slightly harmful
8	60	18	41	2	0	1.74	Slightly harmful
8	80	20	36	0	0	1.64	Slightly harmful
8	100	22	33	0	2	1.68	Slightly harmful

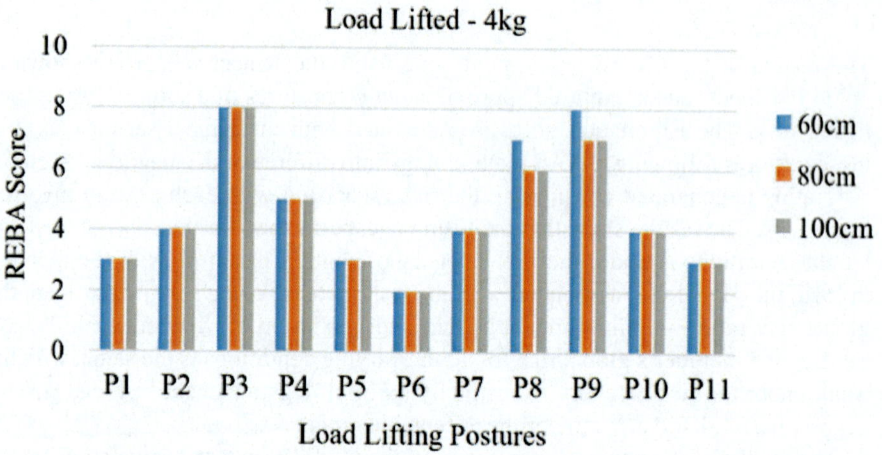

Fig. 4 REBA score rating for postures during lifting of 4 kg load

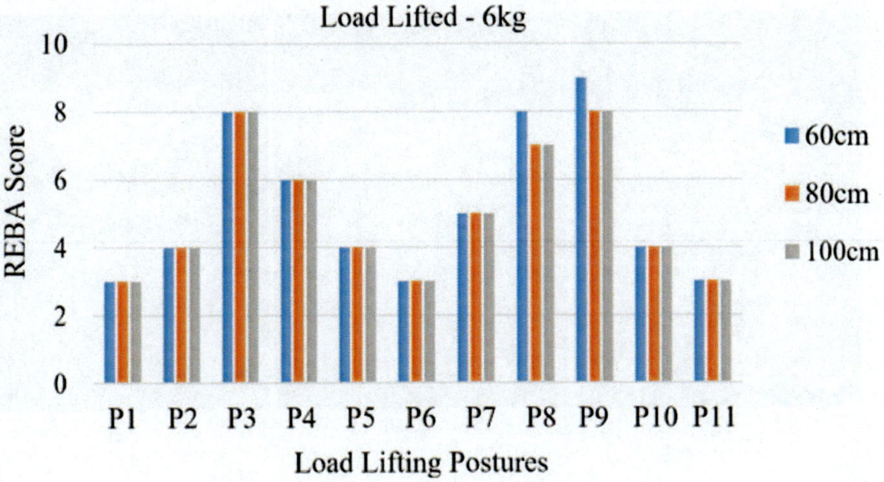

Fig. 5 REBA score rating for postures during lifting of 6 kg load

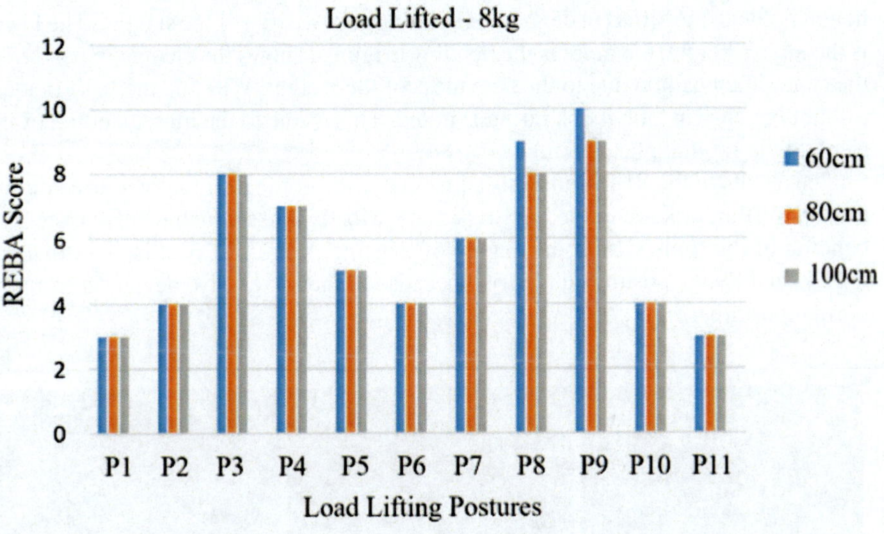

Fig. 6 REBA score rating for postures during lifting of 8 kg load

4 Discussion

The result of the OWAS assessment clearly shows that the performed load-lifting task is under the risk category of slightly harmful. This is because of the twisting and bending of the trunk and sustaining the standing posture during load lifting. Figure 7 shows the postural variation of the trunk and legs due to the change in the destination's height. The change in global risk rating with the change in the table's

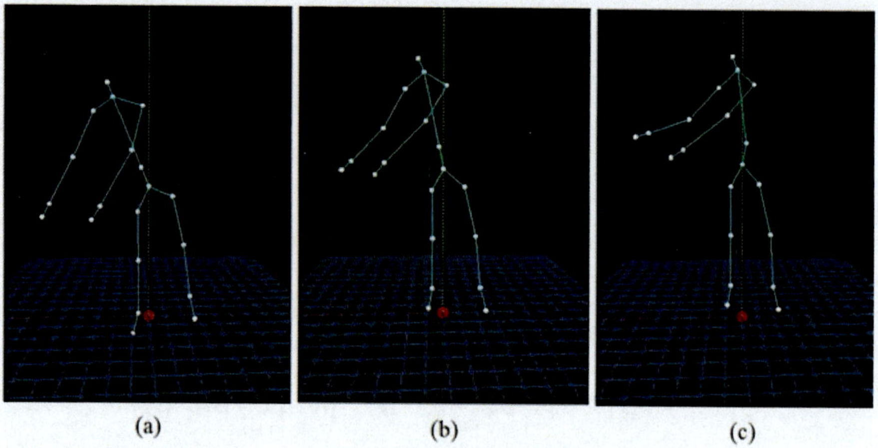

Fig. 7 Skeletal image capture by Kinect, depicting the variation of posture for different table heights. **a** 60 cm, **b** 80 cm, **c** 100 cm

height is due to the different degrees of bending and twisting of the subject. The lower is the table's height, the more is the bending. Figure 8 shows the change in posture in the centerline position due to the heaviness of the weight. With the increase in load, a slight increase in global risk rating is noted. This is due to the more bending of the trunk in the sagittal plane.

The results of the REBA analysis show that some of the positions acquired during the load-lifting task are in the high-risk zone. Mostly those which involve excessive bending of the trunk with some degree of twisting. From Fig. 3, it is clear that P3, P4, P8, and P9 are postures under risk because of their excessive deviation from the normal posture.

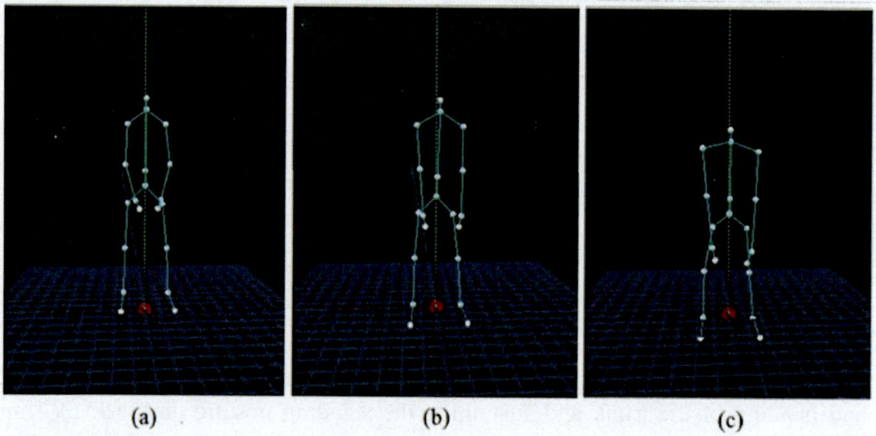

Fig. 8 Skeletal image capture by Kinect, depicting the variation of posture for different load lifted. **a** 4 kg, **b** 6 kg, **c** 8 kg

The postural assessment was done using the REBA, and the OWAS method has a mutual agreement on the results up to a slight extent. The result of both methods agreed on the point that the load-lifting task for the given condition holds a certain level of risk.

However, the assessment carried out using the Kinect sensor holds certain constraints or limitations. To ensure the better accuracy of the posture detection, it is necessary to make continuous eye contact with the Kinect sensor. Moreover, the use of the Kinect sensor is only limited for the short duration study.

5 Conclusion

This experiment is designed carefully to study the discomfort associated with the female body during a load-lifting task. The study has a major application in material handling industries mostly in the unorganized sectors where the females performed load-lifting on a continuous basis and complaint about lower back discomfort. The present research helped in recommending the parameter for the design of the load-lifting task to minimize the discomfort. The study reveals that the level of discomfort during the variation in height is more apparent than that of the load. The main reason for the slight discomfort from both OWAS and REBA analysis can be assigned to the excessive bending of the trunk during lifting a load from 30 cm height. To reduce discomfort in this task, the base height of 30 cm could be increased.

Acknowledgements The authors would like to thank "Ergonautas Lab, Universidad Politecnica de Valencia" for providing the Ergonautas Software and giving license to use it in the assessment of a worker's posture.

References

Butler HL, Kosey JW (2003) The effect of load and posture on load estimations during a simulated lifting task in female workers. Int J Ind Ergon 31(5):331–341

Female labor supply. https://ourworldindata.org/female-labor-supply#female-participation-in-labor-markets-grew-remarkably-in-the-20th-century. Accessed 23 Nov 2018

Dutta T (2012) Evaluation of the Kinect sensor for 3-D kinematic measurement in the workplace. Appl Ergon 43:645–649

Health and safety statistics for manufacturing sectors in Great Britain (2017), http://www.hse.gov.uk/statistics/. Accessed 05 Nov 2018

Hignett S, McAtamney L (2000) Rapid entire body assessment (REBA). Appl Ergon 201–205

Hoozeman MJM, Kigma I, Wiebe HK, Jaap H (2008) Effect of lifting height and load mass on lower back loading. Ergonomics 51:1053–1063

Karwowski W (1991) Psychophysical acceptability and perception of load heaviness by females. J Ergon 34(4):487–496

Liu DR (2014) Comparison of concepts in easy to use methods for MSD risks assessment. Appl Ergon 45:420–427

Mas JAD, Marzal JA (2014) Using the Kinect sensor in observational methods for assessing postures at work. J Appl Ergon 45:976–985

Mital A (1984) Comprehensive maximum acceptable weight of lift database for regular 8-hour work shifts. Ergon 27(11):1127–1138

Mital A (2007) Maximum weights of lift acceptable to male and female industrial workers for extended work shifts. J Ergon 27(11):1115–1126

Mo S, Jung M (2015) Dynamic electromyography analysis of shoulder muscles for one-handed manual material handling. J Ergon Soc Korea 34(5):313–326

World health organization. http://www.who.int/newsroom/factsheets/detail/musculoskeletal-conditions. Accessed 13 Nov 2018

Zurada J (2012) Classifying the risk of work-related low back disorders due to manual material handling tasks. Expert Syst Appl 39(12):11125–11134

Chapter 14
Noise-Induced Hearing Loss and Hand Arms Vibration Syndrome

Balakrishnan Nair

1 Introduction

Noise is an unwanted sound that can be unpleasant, loud or disruptive to hearing (Wikipedia). Worldwide noise is the major cause of all industrial pollutants involving every industry causing hearing losses. Nearly 16% of the hearing loss reported is attributed to occupational noise. According to a World Health Organization fact sheet, updated in Feb 2017, hearing loss posed an annual global cost of $750 billion (Health and Safety Matters 2017). HSE estimates that over 170,000 people in the UK suffer from noise-induced hearing loss as a direct result of excessive exposure at work. Occupational noise can lead to noise-induced hearing loss resulting from long-term exposure to high levels of noise in the workplace. This may lead to temporary or permanent hearing loss. This can further aggravate to lack of concentration, irritation, fatigue, headache, insomnia etc. Ultimately these lead to stress both physically and mentally. In many countries, noise-induced hearing loss is one of the severe occupational hazards and gets compensated. In 2016, harmful noise was officially recognized as an irreversible health hazard.

1.1 Process of Hearing

We hear when sound waves travel through the air to our eardrum, across our middle ear to our inner ear to the auditory center of our brain. Inside the ear, the sound waves change into electrical signals which are carried to the brain by auditory nerves. Our ears continuously carry sound through this hearing pathway. We have two types of hair cells in our cochlea. There are nearly 3500 inner hair cells per ear and about

B. Nair (✉)
HSE Trainer and Consultant, Former Company Representative, Petroleum Development Oman, Muscat, Oman

© The Author(s), under exclusive license to Springer Nature Singapore Pte Ltd. 2022, corrected publication 2022

D. Bijulal et al. (eds.), *Technology-Enabled Work-System Design*, Design Science and Innovation, https://doi.org/10.1007/978-981-16-1884-0_14

12,000 outer hair cells per ear. Inner hair cells collect and transmit relay sound information to the brain through the auditory nerve. Any damage occurring to these hair cells through excessive noise levels causes irreversible hearing loss. Noise levels at workstations shall be mapped to alert employees about noise hazards and their consequences on human health. The process of hearing as detailed here under and the Fig. 1 below should be corelated in conjunction for better understanding.

1. Sound waves enter the outer ear and travel through a narrow passageway called the ear canal or external auditory canal, which leads to the eardrum.
2. Eardrum vibrates from the incoming sound waves and sends these vibrations to three tiny bones in the middle ear.
3. The bones in the middle ear couple the sound vibrations from the air to fluid vibrations in the cochlea.
4. Once the vibrations cause the fluid inside the cochlea to ripple, a traveling wave forms along the hair cells on the basilar membrane.
5. An electric signal is then formed on the auditory nerve which sends signals to the brain which translates into a sound that we recognize and understand.
6. Hearing loss is caused by the damage and death of these hair cells and causes an irreparable effect on hearing.

2 Damaging Effects

World Health Organization estimated nearly 370 million people with hearing deficiencies throughout the world with nearly 91% adults and 9% children. Another interesting factor is that nearly 48–53% of the people with hearing disablement in South-East Asia are above 65 years of age. Though the most common reason for this hearing disablement is noise pollution, 15% of the population is affected by hearing loss. As per the estimates from National Institute on Deafness and Other Communication Diseases, noise-induced hearing loss is one of the most common workplace disorders and the second most self-reported occupational injury. In the USA alone, 30 million workers are at risk due to noise-induced hearing loss and 22 million American adults in the age group of 20–69 are already suffering from it; 44% of carpenters who work with electrically operated wooden saws and 48% of plumbers report hearing loss. Almost closely following in this category are pipe fitters, tile cutters, steel fixers and grinders who work with noisy grinding and drilling machines in their work locations. Overexposure to loud noises over a prolonged period, one may start losing their ability to hear. The damage from this noise exposure is gradual and one may not detect the loss early until you notice that you cannot hear what others say.

In a recent incident, a reputed viola artist was subjected to acoustic shock due to overexposure to noise louder than the permissible level for more than three hours and he suffered a permanent hearing loss while he was in the practice session in the Royal Opera House in London.

Between 2011 and 2014, insurance companies were notified with claims for noise-induced hearing loss and recorded an increase of 189% (Journal and of Otology). However, only 30% of these claims were valid and successful, the remainder was classified as spurious and frivolous. In order to get compensated, claimants should prove their hearing loss as a result of occurrence while they were working. They should also prove their claims that their employers were negligent in discharging their responsibilities in providing a safe workplace and safe equipment.

3 Areas, Occupation and Causes

The construction industry is the most noise-generating field. Various machinery and equipment are required in project works and these produce noise to almost unbearable levels. By exposing to more than recommended and permissible level in the construction works, the employees are subjected to slow and gradual noise-induced hearing loss. Due to very tight delivery schedules of the projects, the working hours may have to be extended from the normal recommended working hours which also are reasons for the workers to overexposure and cross their permissible exposure level to noise at work locations.

Other industries like chemical factories, textile industries, fertilizer plants, oil and gas industries, rail industry, aviation industry, road transport industry etc. are other sectors where employees report noise-induced hearing losses. However, much larger numbers are reported from the construction field. Employees working in thermal power stations and those specifically assigned with operation and maintenance of turbine units are exposed to very severe noise levels and in the long run, if appropriate control measures are not exercised, these employees are at higher risk of losing their hearing ability prematurely and permanently.

Since the damage from noise exposure may be gradual, one may not notice or feel it easily until it becomes prominent and painful. Noise-induced hearing loss can be quick and sudden from sounds of a gunshot or bomb blast or shell explosions which are quite sudden and impulsive in nature. Such sounds can rupture the eardrum causing a permanent damage to the hearing process. Such loud noises can cause Tinnitus (Health and Safety Matters 2017)—a ringing, buzzing, roaring sensation in the ear or head. Tinnitus may subside over time but may continue for a long period.

The permissible unit of exposure level to noise on human ears is 85 dB and any noise level above 85 dB will cause a temporary hearing loss over time. It becomes highly unbearable and the ears are subjected to a ringing sound and may be restored to a normal state within a nominal time period. This temporary hearing loss or dullness of hearing is known as temporary threshold shift (TTS) (Passey et al. 2015a, b, c) which may get resolved within the first 10–15 days of exposure. But this may turn serious if the employees are subjected to repeated exposure to noises above 85 dB for 8 hours for 5 days in a week, and it may develop permanent threshold shift (PTS) (Journal and of Otology). The effect of permanent threshold shift can be devastating and will result in permanent memory loss and psychic disorder. The

Table 1 Noise levels

Equipment/noise generation source	Noise generated (dB)
Gunshot	150
Plane take off	140
Pneumatic drill	130
Chainsaw	120
Football crowd	110
Electric drill	100
Heavy road traffic	90
Hand saw	85
Busy street	70
Normal conversation	60
Average noise	50
Quiet conversation	40

results are irreversible due to the permanent reduction of nerve impulses to the brain and the employee will start developing undue stress and other complications.

Noise Regulations Act provides guidelines for keeping noise at permissible levels so that employees can work with peace and comfort. Noise levels generated from various equipment or workplaces are given in Table 1 (Techniques of Workplace Health and Safety Management).

3.1 *Hands Arm Vibration Syndrome*

Construction equipment and machinery like drilling machines, grinding machines, cutting machines, lawn movers, cranes, welding machines, jackhammers etc. generate vibration during operation. These machines are mostly hand-operated and workmen have to hold these equipments during their work activities. These machines are power-driven either by electrically operated or pneumatically operated and operate at variable speeds of around 1800–3000 rpm. These power tools are hazardous and one should be very careful during working. It also depends upon the usage period. Vibration is felt in the hands of employees when these hand-operated machines are used at work locations. Prolonged usage of such machines causes extreme pain and ultimately affects the dexterity of hand movements leading to loss of sensation.

The risk of developing hands arm vibration depends upon the frequency of vibration generated by the equipment, the exposure period of the worker to the equipment in operation and the tightness of the grip while holding the equipment. Hand-arm vibration develops severe pain in the hands and also causes numbness if exposed to extended usage of a machine or tool. The blood supply to the fingers and the palm gets regulated due to the tightness of the hands while for work holding the machine or tool which is the primary cause for developing vibration white finger (Introduction to Health and Safety at Work). Discoloration or enlargement of the fingers may develop and further deterioration of the fingers can prevent the workmen from being able to hold on to the machines. In more severe cases, gangrene may develop which may lead to even amputation of the affected fingers or hand.

The Control of Vibrations at Work Regulations Act 2005 (Introduction to Health and Safety at Work) details out the work time exposure limit of workmen exposed to vibration. Short-term exposure limits should be followed at any cost to prevent harm to workmen due to its severe effects from vibration. A maximum exposure time of 15 minutes in a working shift of 8 hours is only permitted as per standards and should be strictly followed by employers to keep the employees safe while working with vibration generating equipment or machinery.

Typical high-risk processes where vibration is generated are:

- Grinding, sanding, polishing etc.
- Cutting of stones, tiles, wood, metal etc.
- Riveting, hammering, drilling, bolt tightening by pneumatic wrenches etc.
- Compacting of sand, aggregate or concrete
- Rock breaking, road surfaces cutting and concrete cutting
- Descaling, surface blasting for painting applications.

4 Attenuation (Passey et al. 2015a, b, c)

4.1 Noise

Health and Safety Executive (Health and Safety Executive), who is authorized to conduct inspection at worksite across Britain, recommends that the employers should provide sufficient and adequate measures to protect the employees from any occupational illness including the effects of excess noise on the human ear with appropriate control measures. Awareness about health hazards and ill-health effects due to exposure to noise to all employees through suitable training on different control measures on attenuation techniques. Regular monitoring shall be done by employers about the efficacies of the control measures.

The following measures shall be considered to reduce or control hazards related to noise at the worksite:

- Eliminate the noise-generating source from the worksite, if possible
- Reduction of noise at source by suitable enclosures
- Consider job rotation among employees to reduce exposure time to noise
- Consider lagging the pathway of noise reaching the receiver from the source of generation
- Personal protective equipment like ear defenders or earplugs
- Noise risk assessment and noise surveys at possible noise-generating locations.

The following points shall be included while carrying out a noise risk assessment:

- Details of the equipment or machinery including its condition and maintenance schedule
- Details of noise measuring device with calibration status
- Number of employees working, near and are likely to be affected by noise generation
- Identify the tasks and processes that are likely to generate noise.

Additional measures that may be considered for control and reduction of noise at work locations can be:

- Relocate or reorient the machinery that generates noise to reduce noise levels
- Isolate the machinery by providing enclosures to reduce noise generated
- Identify noise generation zones and prepare a noise map to make employees aware
- Provide information, instruction and training to employees for better awareness
- Use health surveillance schemes to identify employees suffering from hearing losses
- Apply engineering control measures like damping, use of silencers, lagging noisy areas, providing additional enclosures to reduce the noise level, use of absorption walls etc. All these measures are described in "Controlling Workplace Hazards", 2nd Edition of 2015 by Roger Passey, Dr. J.Phelpstead.

4.2 Vibration

Ill-health effects due to vibration shall be controlled using the hierarchy of control measures like that for any high-risk activities. They can be chosen from eliminating the hazards if possible, reducing or substituting the vibration producing equipment, applying insulated equipment to prevent vibration from machine to human, using vibration absorbent gloves and rotating the people after every half hour who are working on vibration generating work activities. Hazards can be reduced by modifying the work processes and practices. Engineering controls shall be applied wherever possible by providing improving equipment or machinery like provision of anti-vibratory pads to reduce vibration. Regular maintenance and service of equipment are important to reduce the wear and tear of machine parts which may generate noise and vibration. The following shall be considered for reducing risks due to vibration:

- Avoid equipment that generates vibration whenever and wherever possible
- Carry out a risk assessment of activity that may cause vibration
- Maintenance schedule of vibratory equipment shall be monitored and adhered
- Identify suitable work measures including job rotation to reduce exposure of workers to vibration
- Provide suitable vibration resistant gloves and other warm clothing to protect vibratory shocks

- Health surveillance should be provided to employees for identification of ill-health and medical treatment
- Provide suitable information, instruction and training to employees about hazards due to vibration
- Consult workers and obtain their views, and concerns for improvement.

5 Conclusion

Noise-induced hearing loss and hands arm vibration syndrome are two very serious occupational hazards that can change the lifestyle of employees. These are the results of prolonged exposures at work locations. Both employers and employees have a combined responsibility toward compliance with noise and vibration regulations acts. Health and Safety Executive, during inspections at work locations in the UK, have exposed employers where workers were found regularly exposed to hand-arm vibration due to working with hand-held power tools. Organizations have been fined and penalized for such violations to effectively change the working attitude and culture. Employers should be more vigilant and take extra care to control the number of workers getting affected by noise-induced hearing loss. This not only affects the life of employees but also costs a lot of money for compensation. Strict regulations, commitment and enforcement plans from employers can make the worker use appropriate wearable technology personal protective equipment. Non-compliance can lead to situations where a large number of workers would be affected by hearing loss. Audiometry checks by trained and competent persons should be conducted on workers exposed to high noise levels. Empower employees to identify noisy sources and identify measures to reduce and control noise and vibration to lead a happy and stress-free life (Fig. 1).

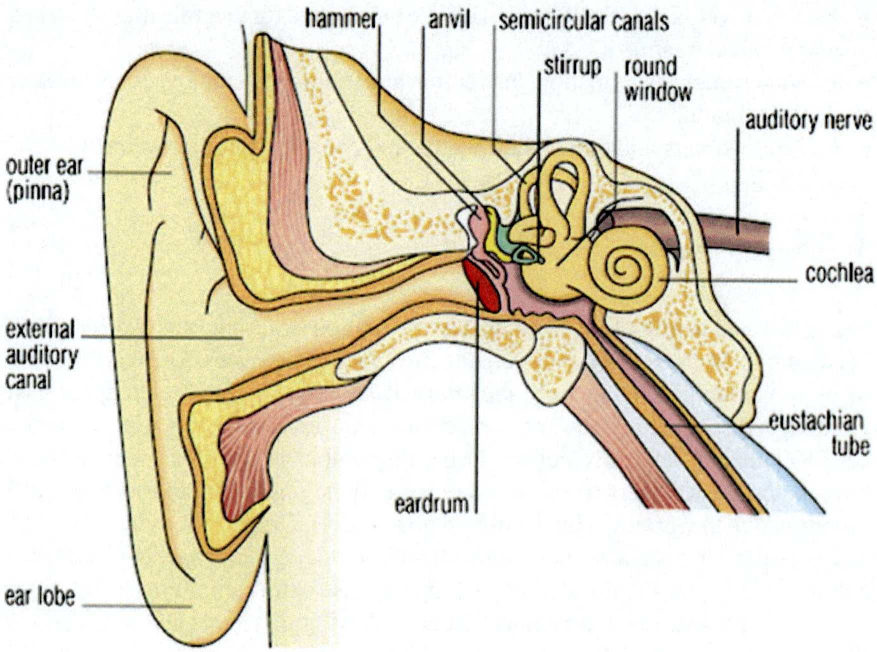

Fig. 1 Human ear (Slideshare.com)

References

Health and Safety Executive, UK Premier Agency for Worksite Inspections

Health and Safety Matters (2017) Trends and innovation in protective equipment and work wear. PPE Insight

Indian J Otol 20(4):2014

Introduction to Health and Safety at Work, Ed Ferret and Paul Hughes, Page 252

Introduction to Health and Safety at Work, Ed Ferret and Paul Hughes, 15.2.3 Page 254

NIH Publication No: 14–4233 Last Updated 07 Feb 2017

Passey R, Phelpstead J, Neasham Z, Towlson D (2015a) Controlling workplace hazards, 2nd edn. Element 8, 8–5

Passey R, Phelpstead J, Neasham Z, Towlson D (2015b) Controlling workplace hazards, 2nd edn. Element 8, 8–3

Passey R, Phelpstead J, Neasham Z, Towlson D (2015c) Controlling workplace hazards, 2nd edn. Element 8, 8–7

Slideshare.com

Techniques of Workplace Health and Safety Management, October 2018 Specification, Page 8

Wikipedia

Chapter 15
Occupational Health Profile of Workers in Terracotta Handicraft Industry: Ergonomics Intervention

Mukesh Kumar Kamti, Nikhil Ghag, Rauf Iqbal, and Karuna Jain

1 Introduction

India is a country of rich culture, history and traditions. It is one of the major producers and suppliers of handicrafts products in the world. India has been a major producer and supplier of handicrafts products for a very long time. Before the industrial development, this art and industry was a potential economic advantage for the country. The Indian handicraft industry is showing a continuous growth rate of 20% every year and it is one of the most important segments of the decentralized sector in India. The Indian handicrafts industry is a highly labour-intensive, cottage-based and decentralized industry. The industry is spread all over the country mainly in rural and urban areas. Most of the manufacturing units are located in rural and small towns, and there is huge market potential in all Indian cities and abroad. The handicraft industry is a major source of income for rural communities employing over six million artisans including a large number of women and people belonging to the weaker sections of the society. The objective of the present study is to estimate the prevalence of musculoskeletal disorders (MSDs) and evaluate the working posture with Rapid Upper Limb Assessment (RULA) and Rapid Entire Body Assessment (REBA) among terracotta handicraft workers. The workers in the handicraft industry suffer from various types of occupational risk factors. Due to the nature of their work and the available workstation resources, workers use floor-sitting postures and are not likely to use chairs. Floor-sitting postures, particularly crossed-leg and heel-sitting postures, are popular among these types of activities. Workers often assume such postures for prolonged periods, which are likely to be one of the major risk factors for low back pain (Solomonow et al. 2003, Keawduangdee et al. 2012).

M. K. Kamti (✉) · N. Ghag · R. Iqbal · K. Jain
National Institute of Industrial Engineering, Mumbai, India
e-mail: MukeshKumar.Kamti.2019@nitie.ac.in

© The Author(s), under exclusive license to Springer Nature Singapore Pte Ltd. 2022, corrected publication 2022

D. Bijulal et al. (eds.), *Technology-Enabled Work-System Design*, Design Science and Innovation, https://doi.org/10.1007/978-981-16-1884-0_15

1.1 Occupational Health Issues of Handicraft Craftsmen

The workers in the handicraft industry suffer from various types of occupational risk factors. Due to the nature of their work and the available workstation resources, workers use floor-sitting postures and are not likely to use chairs. Floor-sitting postures, particularly crossed-leg, back bending and heel-sitting postures, are popular among these types of activities (Figs. 1 and 2). Workers often assume such postures for prolonged periods, which are likely to be one of the major risk factors for low back pain (LBP) (Solomonow et al. 2003, Keawduangdee et al. 2012).

Studies have reported that the workers engaged in the handicraft industry are victims of different occupational disorders and psychosocial stresses. They are victims of headache, hearing problem, skin diseases, lung disorders like silicosis, pain in lower limbs, asthma, backache, joint pains, eye strain, other musculoskeletal disorders and so on (Roy et al. 2010, Purnawati 2007). Studies have reported some of

Fig. 1 Crossed-leg posture

Fig. 2 Back bending posture

the unusual work conditions like unnatural work postures, use of hazardous chemicals, unsafe working practices, long working hours and many risks of work accidents which may increase the risks of health issues (Purnawati 2007).

Seeing the above conditions in the Indian handicraft industry, possible issues arise among Indian terracotta craftsmen:

- Prolonged awkward working posture
- Cross-legged seated posture
- Perceived fatigue and strain
- Improper design of tools
- Improper workstations
- Musculoskeletal disorders

The above-mentioned issues reduce the active work-life of terracotta craftsmen.

Studies have been conducted in the other Indian handicrafts, but none of the studies has looked up into the ergonomic issues in Indian terracotta craftsmen with a proper ergonomic intervention in terms of tool design. There is a lack of ergonomic concepts in developing a new innovative design for the tools used for the handicraft industry.

Considering the above facts, there is a need to study the working conditions in terms of their working pattern, tools used and the workstation they work in. This study would also like to focus on the prevalence of musculoskeletal complaints in the terracotta handicraft industry and to identify the ergonomic risk factors in terms of design gaps present in the tools used for the terracotta handicraft industry. This study also aims to propose an improved design for the tool with ergonomic intervention so that the workers in the handloom industry can work efficiently for an extended period with minimum discomfort.

2 Brief Review of Literature

Mukhopadhya and Srivastava (2010) studied on evaluating ergonomic risk factors in non-regulated stone carving units of Jaipur. He identified different ergonomic risk factors associated with this profession. Objective measurements (heart rate and skin temperature) were recorded with a stopwatch and digital thermometer. The working heart rate after 30 min of work was 112.4 beats per minute categorizing the work as moderately heavy. These indicate the vulnerability of many of the postures to musculoskeletal disorders and injury (Mukhopadhya and Srivastava 2010).

Arphorn et al. (2008) studied on the sculptors' workstation in pottery handicraft to reduce muscular fatigue and discomfort. The improvements of the workstation were redesigned of the banding wheel, storage of carving equipment and adjusted height of seat and results found that discomfort of the general body, left and right low back muscles and right shoulder muscle for operating at the modified workstation were significantly less than operating at the traditional workstation.

It could be summarized that the modified workstation could clearly reduce discomfort in low back muscles and right shoulder muscle with a significant difference. Consequently, there is increased productivity and comfort for the sculptors using this modified workstation (Arphorn 2008). Nurmianto studied ergonomic intervention in handicraft-producing operation. A total of 20 trainees were questioned regarding musculoskeletal disorders (MSDs). Among the trainees, knees, back and shoulders problems were more prevalent compared to other body regions. Based on the problems found, a new workstation was developed. The new workstation improved working posture. Working on the table improved neck, trunk and legs postures. The working posture was improved by developing a new working table (Nurmianto 1987).

Songkham et al. (2008) investigated occupational hazards and health status on 307 pottery workers working at Chiang Mai, Thailand. The major result revealed that the most illnesses among the samples were musculoskeletal disorders including hand-arm-shoulder pain, back pain, neck pain and leg pain. Skin and respiratory tract illnesses found among the samples included skin rash, runny nose, coughing, sneezing and nose irritation (Songkham et al. 2008).

Limmongkon et al. (2008) studied were to assess the effectiveness of the modified sculptors' workstation for reduced muscular fatigue and discomfort in pottery handicraft. The results found that discomfort of body, left and right low back muscles and right shoulder muscle for operating at the modified workstation were significantly less than operating at the traditional workstation.

3 Objectives of the Study

The study was carried out with the following objectives:

- To study various terracotta products and the type of tools/workstations used to develop the products.
- To evaluate the biomechanics of the working posture and postural stress of the terracotta handicraft workers
- To analyse the user interface design between the tools, working condition and the users.
- To study the human factors involved in the terracotta handicrafts operations
- To improvise the existing design of the tools/workstations for the enhanced user interface.

4 Methodology

A cross-sectional study was carried out in the terracotta handicraft industry in Bishnupur, West Bengal. The reasons for the study as well as its goals and contents were communicated to all involved subjects along with how results will be useful for their health improvement. 15 male and 5 female workers in the age ranging from

25 to 60 years were selected, their height and body weight were measured by using standard anthropometric rod and digital weighing machine, respectively. From the above readings, body mass index (BMI) was calculated. The prevalence of MSD was evaluated using Nordic Musculoskeletal Discomfort Questionnaire (NMQ). Detailed photography and techniques like RULA and REBA were used to estimate the posture, force and risk level of the work. The most frequent postures adopted by the workers were considered for the study. Duration of work and frequency of rest break were analysed by using a questionnaire method. A temperature-sensing device was used to find out the temperature of the furnace in the terracotta handicraft industry.

5 Results

Analysis of different working postures that workers most adopted was depicted in Tables 1 and 2. Working posture and activities of terracotta handicraft workers have been analysed by rapid entire body assessment (REBA) and rapid upper limb assessment (RULA) methods. Awkward working postures cause harmful effects on the health or may lead to the development of musculoskeletal disorders. So, the present study shows that the terracotta handicraft workers are at a high risk level, so that is why work-related musculoskeletal disorder (WRMSD) developed and they claimed for pain in their different body parts.

Table 1 Analysis of different working postures by REBA method

S. no	Posture	Activities	REBA score	Risk level	Action to be taken
01		Mixing of clay	11	Very high risk	Implement change
02		Putting clay pot in open sunlight to dry	10	High risk	Investigate and implement change
03		Making the design of clay sculpture	10	High risk	Investigate and implement change
04		Spinning pottery wheel (standing and moving posture)	11	Very high risk	Implement change
05		Keeping clay pot after dry	08	High risk	Investigate and implement change

Table 2 Analysis of different working postures by RULA method

S. no	Posture	Activities	RULA score	Risk level	Action to be taken
01		Spinning the pottery wheel (sitting posture)	05	High risk	Further investigation, change soon
02	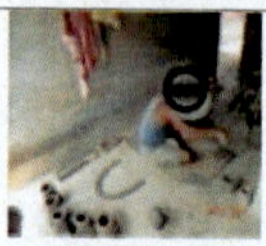	Making small clay materials	03	Low risk	Further investigation, change may be needed
03		Making perfect shape by their tools	07	Very high risk	Investigate and implement change
04		Spinning the pottery wheel by a stick	06	High risk	Further investigation, change soon

Besides this, the handicraft workers always do work in floor sitting and stooping postures for a long time without taking any break and this may be another reason for their musculoskeletal disorders (MSDs). During their activities, they always do the movement of wrist, hands and back that may cause their shoulder, neck, wrist and back pain more.

In Tables 1 and 2 analysis of different working postures of terracotta handicraft workers was done by REBA and RULA method and it has been found that they are at high risk level. That is why workers are suffering from work-related musculoskeletal problems and are claimed for pain in their different body parts. During clay mixing and spinning of the pottery wheel, the REBA score is 11, which indicates a very high risk level.

Table 3 indicates the distribution of the subjects in different categories according to explanatory variables like gender, age, BMI, working hour/day and rest break frequency. Table 4 indicates the prevalence of musculoskeletal disorder (MSD) in body regions like neck (90%), shoulder (95%), upper back (75%), lower back (90%), wrist (80%), forearm (90%) and knees (85%).

Measurement of grip strength is an important component for hand efficiency (Simmonds et al. 2015). In this study we found that the mean grip strength in kg [right hand (32.26 ± 4.38), left hand (35.46 ± 7.66)] for the terracotta handicraft workers and [right hand (31.4 ± 7.28), left hand (31.08 ± 8.96)] for the control group. The right-hand grip strength has been found lower in terracotta handicraft workers (Table 5).

In this study after comparing with the previous experimental results among different occupational workers, it was found that there were differences in the level

Table 3 Distribution of the subjects in different categories according to explanatory variables

Explanatory variables	Number of subjects (N = 20)	Percentage (%)
Male	15	75
Female	5	25
Age in years		
25–30	04	20
31–35	10	50
36–40	03	15
41–45	02	10
>45	01	05
BMI		
<18	02	10
18–25	07	35
26–30	08	40
<30	03	15
Working h/day		
> 6 h	03	15
7–8 h	12	60
>8 h	05	25
Rest break frequency		
Once in < 2 h	11	55
2–4 h	03	15
>4 h	07	35

Table 4 Prevalence of MSD in different body regions of terracotta handicraft workers (N = 20)

Body regions	Number of subjects (N = 20)	Percentage (%)
Neck	18	90
Shoulders	19	95
Upper back	15	75
Lower back	18	90
Wrist	16	80
Forearm	18	90
Knees	17	85

of maximal static forces as well in the level of muscular fitness among the different occupational workers. The observed values of handgrip strength showed that the terracotta handicraft industry workers were highly affected in comparison with the other occupational groups (Table 6).

Table 5 Handgrip strength in kg

Handgrip strength	Group	Handgrip strength (kg)
		Mean ± SD
Right hand	Terracotta handicraft workers	33.26 ± 4.38
	Control group	35.46 ± 7.66
Left hand	Terracotta handicraft workers	31.4 ± 7.28
	Control group	31.08 ± 8.96

Table 6 Handgrip strength of other occupational groups

Handgrip strength	Group	Handgrip strength (kg)
		Mean ± SD
Right hand	VDT operators	43.6 ± 5.9
	Industrial workers	42.6 ± 4.1
	Safety inspectors	48.4 ± 5.5
Left hand	VDT operators	42.8 ± 7.1
	Industrial workers	42.9 ± 5.6
	Safety inspectors	45.2 ± 4.0

The maximal forces exerted by humans are important in many aspects of life, mainly in occupational work. People whose occupations involve spending prolonged periods at an improperly designed workstation are exposed to muscular stress that may lead to the development of MSDs.

Similar to handgrip strength values the mean values of three types of pinch (tip, key and palmer) strength were found different between workers and the control group. Mean tip pinch strength of right and left hand was found to be 3.69 ± 0.94 and 3.31 ± 1.0, respectively, for terracotta workers, which is lower than the control group (right = 4.05 ± 1.34 and left = 3.43 ± 0.96) (Table 7).

Table 7 Pinch strength in kg

Pinch strength		Group	Pinch strength (kg)
			Mean ± SD
Tip pinch	Right	Terracotta handicraft workers	3.69 ± 0.94
		Control group	4.05 ± 1.34
	Left	Terracotta handicraft workers	3.31 ± 1.0
		Control group	3.43 ± 0.96
Key pinch	Right	Terracotta handicraft workers	8.52 ± 1.38
		Control group	7.54 ± 1.39
	Left	Terracotta handicraft workers	7.61 ± 1.73
		Control group	7.02 ± 1.79
Palmer pinch	Right	Terracotta handicraft workers	5.40 ± 1.24
		Control group	5.32 ± 1.49
	Left	Terracotta handicraft workers	4.43 ± 1.65
		Control group	4.32 ± 1.24

6 Discussion

The result of this study revealed that the terracotta handicraft workers are engaged in floor-sitting postures, particularly crossed-leg and heel-sitting postures in their working condition. There is a need to redesign the work/workstation and improve working posture. The study showed that more than 65% of the average values of the subjects suffered from work-related musculoskeletal problems. The study carried out by Montreuils et al. (1996) on textile tufting workers handling thread cone has reported that 64.9% of workers had at least one work-related musculoskeletal pain among various sites in body regions. The research outcome of the study conducted by Punnett et al (1985). related to female garment workers indicates that the majority of participating subjects in the study are experiencing shoulder, back and wrist pain because of the repetitive nature of the operation and a poor design of the spinning wheel. Grandjean (1982) reported the disadvantage of sitting posture which causes a disturbance in digestion and breathing. A prolonged slaking of the abdominal musculature and the purported ill effect of the flexion of the lumbar spine also contribute to this. This study also reflected the working posture analysis using RULA and REBA. Identical research works conducted by Ghosh et al. signify findings related to musculoskeletal pains due to awkward posture adopted by the goldsmiths in India.

7 Conclusion

Through this study it has been found that the health of terracotta handicraft workers is affected and due to very high degree of risk these workers would develop MSDs in various regions of the body. Results indicate to improve the design of workstation and the working posture for terracotta workers. The majority of the subjects in this industry had shoulder, neck, back, knee and wrist pain due to the repetitive nature of the job and the improper design of the workstation and pottery wheel. MSDs are prominent in this study due to prolonged sitting postures, particularly crossed-leg and heel-sitting postures without any backrest and remain in the position for at least 8 h a day. As there are residual pain and fatigue, the productivity slows down. The RULA (7) and REBA (11) scores indicate that the posture of the workers is at very high risk level and needs urgent investigation and implement change for modification. Through temperature-sensing device it has been found that the temperature (600 °C) of the furnace is not sufficient for making terracotta products. So, there is a need to modify the design of the furnace to make good products. Dynamic and repetitive loads experienced by the workers lead to the prevalence of MSDs in the operators. Perceived influence indicates the urgent need of redesigning the workstation and pottery wheel to reduce the prevalence of MSDs in terracotta handicraft workers.

References

Arphorn S, Limmongkok Y, Pruktharathikul V, Jalayondeja W, Chaikittiporn C (2008) The modified sculptors workstation in pottery handicraft for reducing muscular fatigue and discomfort. The first east asian ergonomics federation symposium. Japan, pp 12–14

Grandjean E (1982) Fitting the task to the man: an ergonomic approach. Taylor and Francis, London

Limmongkon Y, Arphorn S, Chaikittiporn C, Pruktharathikul V, Singhakajen V, Jalayondeja W (2008) Effectiveness of modified sculptors' workstation for reduced muscular fatigue and discomfort in pottery handicraft, occupational hazards and health status among pottery workers in chiang mai province, Thailand. 3rd International scientific conference on occupational and environmental health, organized by Vietnam association of occupational health national institute of occupational and environmental health in collaboration with the university of Washington. USA, pp 155

Montreuils S, Laflamme L, Pellier C (1996) Profile of the musculo-skeletal pain suffered by textile tufting workers handling thread comes according to work, age and employment duration. J App Ergon 27:85–91

Mukhopadhyay P, Srivastava S (2010) Ergonomics risk factors in some craft sectors of Jaipur. The Off J Hum Fact Ergon Soc Aust 4–14

Nurmianto E (1987) Ergonomic intervention in handicraft producing operation. Ind Eng 1008–1011

Punnett L, Robins JM, Wegman DH, Keyserling WM (1985) Soft tissue disorders in the upper limbs of female garment workers. Scand J Work Environ Health 11:417–425

Purnawati S (2007) Occupational health and safety-ergonomics improvement as a corporate responsibility of a Bali handicraft company: a case study. J Hum Ergol 36(2):75–80

Roy AK, Medhi JK, Baruah R, Samanta A, Gupta R (2010) Environment, occupational health and safety in the craft sector in India, Base line study of selected craft clusters. All india artisans and craft workers welfare association

Sharma TP, Borthakur SK (2010) Traditional handloom and handicrafts of Sikkim. Indian J Trad Knowl 9(April):375–377

Songkham W, Chanpraist C, Kaewtammanukul T (2008) Occupational hazards and health status among pottery workers in chiang mai province, Thailand. In: 3rd International scientific conference on occupational and environmental health, organized by Vietnam association of occupational health national institute of occupational and environmental health in collaboration with the university of Washington. USA, pp 59

Chapter 16
Assessment of Ergonomic Risk for Work-Related Musculoskeletal Disorders Among Foundry Workers

Asif Qureshi and K. Manivannan

1 Introduction

Work-related musculoskeletal disorders (WMSDs) occur when the physical capabilities of the worker do not match the physical requirements of the job (Tayyari and Smith 1997). Prolonged exposures to ergonomic risk factors can cause damage to a worker's body and lead to MSDs. Awkward postures, repetitive motions and load handling are significant causes of MSDs at the workplace (Basahel 2015; Fernandes et al. 2011; Nejad et al. 2013; Nimbarte 2014; Parida and Ray 2015).

Ergonomists focus on the identification, quantification and estimation of ergonomic risk through assessment tools using the direct, semi-direct and indirect methods. Direct methods include the application of sophisticated electronic devices and sensors on a human body to measure work postures. The real-time application of these devices in the working condition and higher operating cost generally impede the application of direct methods. Alternatively, semi-direct and indirect methods are applied to quantify the ergonomic risks. Semi-direct methods include the computer programs enabling human posture evaluations through video recording and photographs. Semi-direct methods are broadly classified based on MSD into three classes, viz.: (a) repetitive movements, (b) strained postures and (c) handling of loads. The popular techniques of repetitive movements are RULA, Job strain index, IBV, OCRA, while REBA, OWAS, Vira, PATH methods are applied to evaluate MSDs due to strained postures. The techniques such as NIOH, KIM and MAC are used to assess the ergonomic risks due to load handling (Buckle 2005; Crawford et al. 2008; Gómez-Galán et al. 2017; Valero et al. 2016).

Indirect methods use the subjective assessment of operators through standardized questionnaires. Some of the popular questionnaires are Nordic, Keyserling, Quick

A. Qureshi (✉)
KIT College of Engineering Kolhapur, Kolhapur, India

A. Qureshi · K. Manivannan
Vellore Institute of Technology (VIT) Vellore, Vellore, India

© The Author(s), under exclusive license to Springer Nature Singapore Pte Ltd. 2022, corrected publication 2022

D. Bijulal et al. (eds.), *Technology-Enabled Work-System Design*, Design Science and Innovation, https://doi.org/10.1007/978-981-16-1884-0_16

Exposure Check (QEC). These methods sometimes need to be validated statistically and limited to the views of a sample population only (Gómez-Galán et al. 2017; Valero et al. 2016).

The classical ergonomics research presents the tussle between semi-direct and indirect methods. Factors related to WMSD are complex and therefore researchers argue that there are no specific guidelines for the selection of method. Chiasson et al. compared eight methods used to evaluate risk factors associated with MSD and concluded that no two methods are in perfect agreement (Chiasson et al. 2012). Roman-Liu compared the methods for assessing external load causing MSD and concluded that it is necessary to develop a comprehensive method appropriate to all body parts and all work tasks (Roman-Liu 2014). Waters et al. highlight the complexity in the evaluation of risks of MSD and suggest step by step procedure to select suitable tools for risk assessment (Waters et al. 2016). Kee et al. compared the results of OWAS, RULA and REBA and did not find any correlation between them. OWAS and REBA underestimated the results than RULA (Kee et al. 2016). Jones et al. compared five postural methods and found moderate agreement between the methods and the results varied with jobs (Jones et al. 2016).

The proponents of semi-direct methods argue that semi-direct methods are precise and capture real-time workspace issues, while indirect methods are quick, comprehensive and researchers' bias-free. The risk output of each method depends on exposures (exertion, posture, repetition, load etc.) and their magnitude considered. As the weightages assigned are different for each method, agreement between methods changes with the job profile. Thus, the reliance on the single paradigm either semi-direct or indirect is inadequate to explain the ergonomics risks causing WMSDs.

The aim of this paper is to assess the ergonomic risk for WMSDs in foundry worksystems using both semi-direct (REBA) and indirect (QEC) methods. Further, we compare the results of these two techniques and comment on the effective comprehension of techniques in ergonomic risk assessment.

We specifically choose the foundry worksystem as the context for this study due to the following reasons: (i) Foundry worksystems are typically characterized by forceful exertions, repetitive work cycles, awkward postures and whole-body vibrations which are the risk factors associated with MSDs (Armstrong et al. 2002; Ilangkumaran et al. 2014). (ii) Foundry worksystem involves a significant amount of MMH activities, leading to further aggregation of MSD prevalence. (iii) Dearth of foundry-specific studies considering posture analysis.

2 Methodology

In this study, we compare semi-direct (REBA) and indirect (QEC) methods to assess the prevalence of WMSDs in the foundry worksystems. Specifically, REBA is selected for this purpose as foundry worksystems are characterized by awkward work postures and excessive manual material handling. Thus, REBA as an effective strained posture analysis is applied to quantify the risk of MSD. Additionally, we use QEC that allows physical work activities to be assessed in collaboration with the worker. We further compare the results of both REBA and QEC with the correlation analysis.

2.1 Subjects

We included 105 workers as subjects from nine foundries in Western India. These foundries are small-scale units with manual material handling. All the subjects are male, ranging from 20 to 52 years of age. The consent of management and workers was obtained before the study. The subjects selected belong to the worker and helper category from four major departments: fettling, melting, molding and pattern making. The study was carried out during working hours of the day shift (i.e. between 8 a.m. and 5 p.m.).

2.2 Data Collection

For REBA analysis, photographs of 105 workers in selected working postures were taken from different angles. Care was taken to obtain angles of body parts accurately.

The standard analysis procedure for sections A and B of REBA was adopted for examining respective body parts (neck, trunk, leg, arm and wrist). The angles were measured by marking lines along the relevant body segments on the photographs (Fig. 1). REBA scores were calculated and risk levels evaluated for each activity department-wise (pattern making, molding, melting-pouring and fettling).

QEC analysis is a combination of observer's assessment and worker's assessment. The interviews of the same 105 workers were taken as per standard QEC protocol. Observations were made to complete the observer's assessment.

We use data inputs from REBA and QEC to derive scores and action levels based on the ergonomic risk involved. Further, we use correlation analysis to present the comparative of REBA and QEC for both scores and action levels using Minitab®-16.

3 Results

A total of 105 workers were assessed for risk of WSMD from four departments of foundries, viz., fettling, molding, melting and pouring, and pattern making. Table 1 summarizes the demographic details of the total study population. The mean age of the study population is 31.48 (8.45) years with a mean height of 163.29 cm.

The REBA score varies from low (2) to very high risk level (11) among the total sample. As no subject has a REBA score of 1, which is a negligible risk level, this level is not considered for analysis. 30.47% of the sample is exposed to high risk

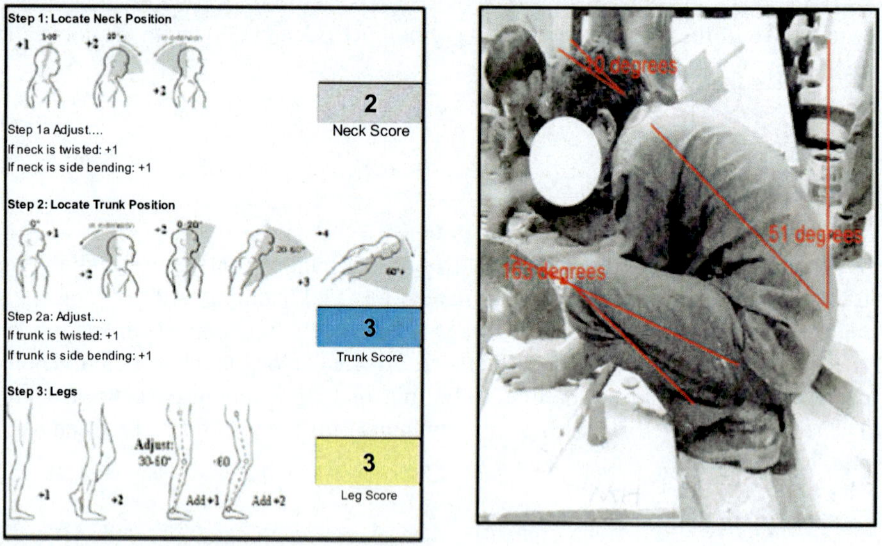

Fig. 1 Illustration of REBA score calculation using photograph

Table 1 Demographic details of a sample population

No	Particulars	Mean (SD)	Minimum	Maximum
1	Age (years)	31.48(8.45)	19	61
2	Weight (kg)	57.63(7.04)	40	82
3	Height (cm)	163.29(6.47)	136	188
4	Foundry experience (years)	7.25(4.90)	1	25

level and 16.19% to very high risk level (Table 2 and Fig. 2). Thus, activities carried out by the above said 46.66% of the workers are risky and need immediate action. Population exposed to medium risk level is also considerable (44.76%) and action is necessary for those.

Department-wise risk analysis reveals that workers from the melting and pouring departments are prone to very high risk levels (44%), followed by workers from the fettling department (13.33%) (Table 2 and Fig. 3).

According to the QEC score, four standard action levels—low, moderate, high and very high—are decided. Out of the total study population, 75.23% fall under high risk level, whereas 11.42% study population falls under very high risk level (Table 2 and Fig. 4). Department-wise analysis indicates that the melting department has the highest percent (20%) of very high risk level and the fettling department has the highest percentage of high risk level (Table 2 and Fig. 5).

The body part wise QEC analysis reveals that from the total population, the mean score of risk for the back (moving) is 30.76 (high), for shoulder/arm is 32.11(high), for wrist/hand is 28.97 (moderate) and for the neck is 14.10 (very high).

Table 2 Action levels for both REBA and QEC

Department(n)	ACTION LEVEL							
	Low n (%)		Moderate n (%)		High n (%)		Very high n (%)	
	REBA	QEC	REBA	QEC	REBA	QEC	REBA	QEC
Fettling (30)	1(3.33)	0	16(53.33)	0	9(30)	29(96.66)	4 (13.33)	1(3.33)
Melting and pouring (25)	1(4)	2(8)	7(28)	2(8)	6(24)	16(64)	**11(44)**	**5(20)**
Molding (41)	5(12.2)	1(2.22)	21(51.21)	7(17.07)	14(34.14)	28(68.29)	1(2.22)	5(12.2)
Pattern making (09)	2(22.22)	0	3(33.33)	2(22.22)	3(33.33)	6(66.66)	1(11.11)	1(11.11)
Total (105)	9(8.57)	3(2.85)	47(44.76)	11(10.47)	32(30.47)	79(75.23)	17(16.19)	12(11.42)

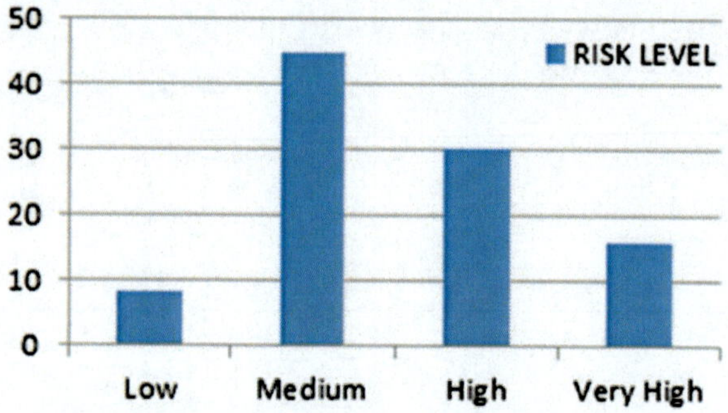

Fig. 2 Percentage risk levels by REBA of the study population (n = 105)

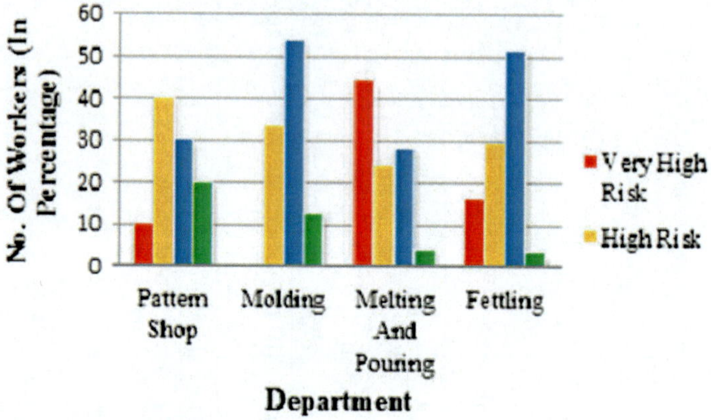

Fig. 3 Department-wise REBA risk analysis

Fig. 4 Percentage risk levels by QEC of the study population (n = 105)

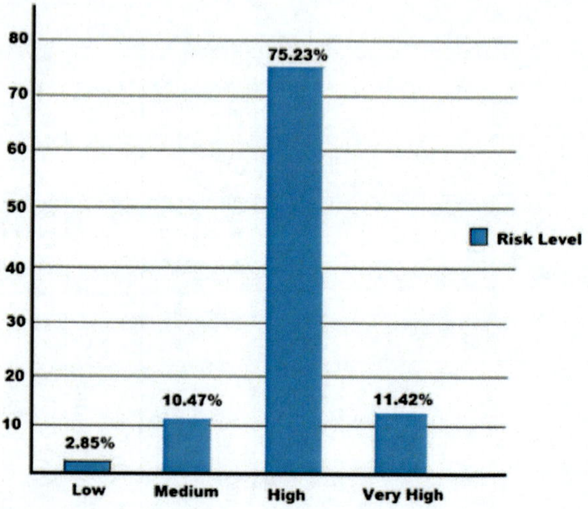

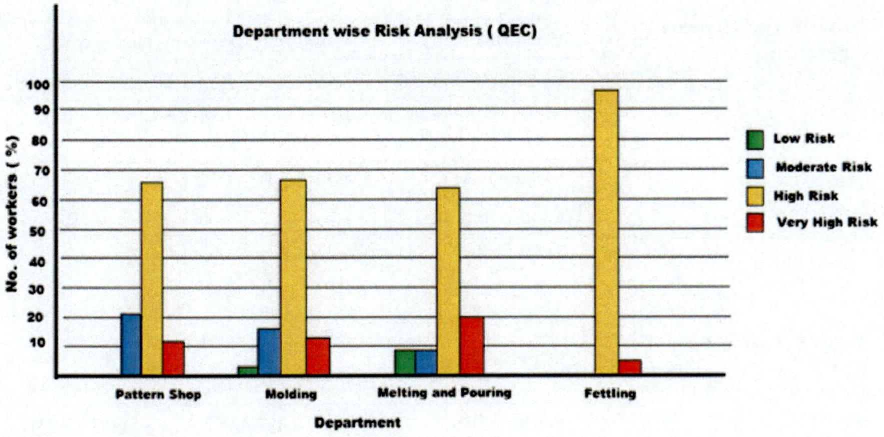

Fig. 5 Department-wise QEC risk analysis

Department-wise QEC analysis for body part was carried and the back and shoulder mean score is highest for the melting department (36.16—high and 34.48—high, respectively). The Wrist score is maximum for the fettling department (30.8—high) and the neck score is maximum for pattern making (16.66—high) (Table 3).

In QEC analysis other factors considered are driving, vibration, working pace and stress. For the study population considered, driving is not applicable and neglected in this analysis. Table 4 gives the department-wise mean score of these factors. QEC score for vibration parameter is maximum (7.16) for the fettling department. For the melting department, both working pace score as well as stress score are maximum (4.84 and 4.44, respectively).

A comparison of REBA and QEC analysis is given in Table 5. From both analyses, the mean risk score for the melting department is the highest followed by the fettling department.

Statistical analysis was done to check the correlation between REBA and QEC outcomes. The Pearson's correlation coefficient for REBA and QEC score is 0.219

Table 3 Body part wise QEC scores

Department (n)	Back mean (SD)	Shoulder/Arm mean (SD)	Wrist/Hand mean (SD)	Neck mean (SD)	QEC score mean (SD)
Fettling (30)	29.2(3.91)	31.2(5.47)	30.8(3.26)	15.73(1.01)	60.75(6.04)
Melting and pouring (25)	36.16(8.50)	34.48(4.25)	29.04(3.00)	12.24(2.18)	63.59(6.04)
Molding (41)	29.26(6.14)	31.80(5.41)	27.80(5.24)	13.58(3.00)	58.21(8.77)
Pattern making (09)	27.77(4.84)	30(5.65)	27.55(4.66)	16.66(2.00)	57.95(7.64)
Total (105)	30.76(6.82)	32.11(5.31)	28.97(4.33)	14.10(2.75)	60.2(7.95)

Table 4 QEC scores for other factors

Department (n)	Vibration	Working pace	Stress
Fettling (30)	7.16(3.18)	4.03(1.62)	3.66(2.91)
Melting and pouring (25)	1(0)	4.84(2.57)	4.44(2.25)
Molding (41)	4.73(3.49)	3.12(1.38)	3.73(2.39)
Pattern making (09)	1(0)	1.66(1.32)	1(0)
Total (105)	4.21(3.67)	3.66(1.99)	3.64(2.55)

Table 5 Comparison of REBA and QEC scores

Department (n)	REBA score mean (SD)	QEC score mean(SD)
Fettling(30)	7(2.43)	60.75(6.04)
Melting and pouring (25)	**8.56(3.4)**	**63.59(7.84)**
Molding (41)	6.65(2.52)	58.21(8.77)
Pattern making (09)	6.55(3.28)	57.95(7.64)
Total (105)	7.2(2.86)	60.89(7.95)

with $p = 0.025$ (<0.05). The probability curve was plotted to insight the relation between action levels of the two methods.

4 Discussions

The major outcome of this study is the presence of a high risk of WMSD in foundry workers. REBA analysis indicates that out of the total population 30.47% is under high risk and 16.19% is under very high risk. However, QEC output reveals that 75.23% is under high risk and 11.42% is under very high risk. Due to higher percentage of very high and high risk categories in both REBA and QEC analysis, immediate interventions are necessary. Administrative and engineering controls are a must in the majority of activities.

The department-wise analysis indicates that the highest percentage of very high risk activities are from the melting and pouring department for both REBA and QEC, followed by the fettling and molding departments. Activities like pouring molten metal, fettling and preparing mold are found to be more prone to MSDs due to excessive load handled, awkward posture or a combination of both.

QEC analysis specifies body part wise risk which is not possible in REBA. The result of QEC analysis of the total population indicates a high score of risk exposures for shoulder (32.11), back (30.76) and neck (14.10). For the melting and pouring department both back and shoulder scores are highest, 36.16 and 34.48, respectively. This is due to the fact that the melting and pouring department has more manual material handling activities with awkward postures than any other department which may lead to risk of WMSD. The fettling department has the highest score for wrist/hand (30.8) which is because fettling activity involves awkward wrist and hand postures with load. The pattern making department has the highest score of neck (16.66) because this department involves the majority of activities with a bent neck.

The QEC score for vibration parameter is maximum (7.16) for the fettling department because workers from this department are mainly working with grinders, pneumatic hammers, etc. In the melting department, workers need to pour the molten metal before it cools down below a specific temperature in a particular number of molds. Due to this both working pace score as well as stress score are maximum (4.84 and 4.44, respectively) for the melting and pouring department.

Both REBA and QEC outcomes are similar and reveal that the melting and pouring department activities are at very high risk. For QEC and REBA scores, Pearson's correlation coefficient is 0.219 with $p = 0.025$ (<0.05). This indicates there is a moderate correlation between the scores of the two methods. This result is similar to a study in different sectors by Chiasson et al. (coefficient 0.35) (Chiasson et al. 2012).

Even though there is a moderate correlation between REBA and QEC scores, in this study, there is no strong relationship between action levels. QEC overestimates the risk than REBA. The overall percentage of high risk by QEC is 75.23%, whereas by REBA is 30.47%. This is due to the fact that in QEC percentage of moderate risk level is shifted to high risk due to manual material handling in foundry activities. The overall percentage of medium risk by QEC is only 10.47%, whereas by REBA is 44.76%. Therefore the probability curve for the QEC action level is steeper than the REBA action level (Fig. 6).

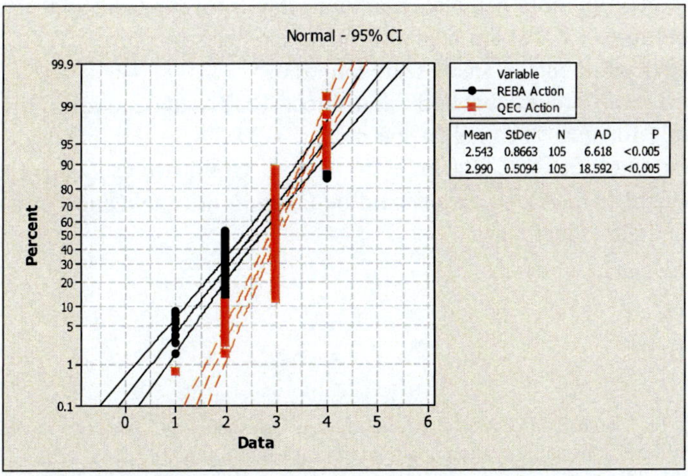

Fig. 6 Probability distribution of REBA and QEC action levels

Indirect methods (QEC in this research) are quick and easy but a subjective response of indirect method may lead to overestimation of risk. To get comprehensive realistic insights, the combined use of the semi-direct method (REBA in this research) and indirect method as used in this study is preferable.

Majid Motamedzade et al. compared the results of REBA and QEC in Engine Oil Company and found a strong relationship between both scores and action levels (Motamedzade et al. 2011). This result differs from the present study due to the changed context. Small-scale foundry worksystem involves more material handling than Engine Oil Company.

5 Conclusion

There is a dearth of studies related to the risk of WMSD in the foundry context, particularly using postural analysis tools. This study reveals that workers from small-scale foundries are at high risk of WMSDs and need immediate attention. Melting and pouring activities have a higher risk of WMSDs compared to other departments. Thus, designing and developing the interventions for the melting and pouring section is an apt extension to this work. Molten metal pouring ladle as an engineering intervention is under development focusing on the reduction of WMSDs risks.

Secondly, the study results show back and shoulder are more susceptible body parts to the risk of MSD in small-scale foundries due to higher manual material handling. These observations can be used to prioritize engineering, educational and enforcement interventions in small-scale foundries considering the WMSDs injury data. Increment in mold heights, ergonomic stands for the ladle and standard operating procedures (SOPs) are taken as the action tasks to address WMSDs risks in concerned departments of small-scale foundries.

The results also reveal that both semi-direct (REBA) and indirect (QEC) methods are suitable for quantification of the risk of WMSDs in the foundry context and reveal a similar outcome. But indirect method (QEC) overestimates risk than the direct method (REBA), so care should be taken in the application of a mix of the indirect and direct methods.

References

Armstrong TJ, Marshall MM, Martin BJ, Foulke JA, Grieshaber DC, Malone G (2002) Exposure to forceful exertions and vibration in a foundry. Int J Ind Ergon 30(3):163–179

Basahel AM (2015) Investigation of work-related musculoskeletal disorders (MSDs) in warehouse workers in Saudi Arabia. Proced Manuf 3(Ahfe):4643–4649, Elsevier B.V

Buckle P (2005) Ergonomics and musculoskeletal disorders: overview. Occup Med 55(3):164–167

Chiasson M, Imbeau D, Aubry K, Delisle A (2012) Comparing the results of eight methods used to evaluate risk factors associated with musculoskeletal disorders. Int J Ind Ergon 42(5):478–488, Elsevier Ltd

Crawford JO, Laiou E, Spurgeon A, McMillan G (2008) Musculoskeletal disorders within the telecommunications sector-a systematic review. Int J Ind Ergon 38(1):56–72

Fernandes RDCP, Carvalho FM, Assunção AÁ (2011) Prevalence of musculoskeletal disorders among plastics industry workers. Cad Saude Publica 27(1):78–86

Gómez-Galán M, Pérez-Alonso J, Callejón-Ferre Á-J, López-Martínez J (2017) Musculoskeletal disorders: OWAS review. Ind Health 55(4):314–337

Ilangkumaran M, Karthikeyan M, Ramachandran T, Boopathiraja M, Kirubakaran B (2014) Risk analysis and warning rate of hot environment for foundry industry using hybrid MCDM technique. Saf Sci 72:133–143 Elsevier Ltd

Jones T, Kumar S, Jones T (2016) Comparison of ergonomic risk assessment output in four sawmill jobs comparison of ergonomic risk assessment output in four sawmill jobs. 3548. at:https://doi.org/10.1080/10803548.2010.11076834

Kee D, Karwowski W, Kee D, Karwowski W (2016) A comparison of three observational techniques for assessing postural loads in industry a comparison of three observational techniques for assessing postural loads in industry. 3548. https://doi.org/10.1080/10803548.2007.11076704

Motamedzade M, Ashuri MR, Rostam Golmohammadi HM (2011) Comparison of ergonomic risk assessment outputs from rapid entire body assessment and quick exposure check in an engine oil company. J Res Health Sci 11(1):26–32

Nejad NH, Choobineh A, Rahimifard H, Haidari HR, Reza Tabatabaei SH (2013) Musculoskeletal risk assessment in small furniture manufacturing workshops. Int J Occup Saf Ergon 19(2):275–284

Nimbarte AD (2014) Risk of neck musculoskeletal disorders among males and females in lifting exertions. Int J Ind Ergon 44(2):253–259

Parida R, Ray PK (2015) Factors influencing construction ergonomic performance in India. Proced Manuf 3:6587–6592

Roman-Liu D (2014) Comparison of concepts in easy-to-use methods for MSD risk assessment. Appl Ergon 45(3):420–427, Elsevier Ltd

Tayyari F, Smith J (1997) Occupational ergonomics principles and applications, 1st edn. Springer, US

Valero E, Sivanathan A, Bosché F, Abdel-Wahab M (2016) Musculoskeletal disorders in construction: a review and a novel system for activity tracking with body area network. Appl Ergon 54:120–130

Waters TR, Lu M, Occhipinti E (2016) New procedure for assessing sequential manual lifting jobs using the revised NIOSH lifting equation new procedure for assessing sequential manual lifting jobs using the revised NIOSH lifting equation. 139. at:https://doi.org/10.1080/00140130701674364

Chapter 17
Risks of Developing WMSDs in Buffing Task: A Case Study

Nusky Ishrat and Abid Ali Khan

1 Introduction

In a developing country like India, where workers consider the harsh working environment as a part of their job, many of them continue to work in such adverse conditions for a major part of their life. WMSDs are a matter of concern in these industries because they can affect the competitiveness of workers in terms of their compensation, labor turnover, reduced productivity, and poor quality of work (Andersson 1992). There is an increasing need to improve workers' safety, productivity, and manufacturing quality. The most common drawbacks of the unorganized sector and local industries are improper workplace design, jobs are not well structured, workers' abilities do not match with job requirements, a hostile working environment, improper working postures, and inappropriate management programs. These problems lead to workplace injuries, mechanical equipment accidents, disabilities, deteriorate health of workers, WMSDs, and over a period of time reduce the productivity of workers and quality of product and hence increase product cost.

Repetitive movements and extremely awkward postures could increase the risk of WMSDs. Hence, cost-effective measurement and tools to quantify the extent of physical exposure to these adverse working postures are of paramount importance (Andrews et. al. 1998). In literature, various postural analysis methods are available to find and measure postural stress during the job. These include posture analysis by (Corlett and Bishop 1976), OWAS (Karhu et al. 1977), RULA (McAtamney and Corlett 1993), PATH (Buchholz et al. 1996), REBA (Hignett and McAtamney 2000), etc. The methods used to quantify postural stresses can be divided into two basic categories: observational and instrument-based techniques. In the present study observational technique has been used as it is more prevalent in the industry, does not cause any interference to the worker during observations, and also does not demand

N. Ishrat (✉) · A. A. Khan
Aligarh Muslim University, Aligarh, India

© The Author(s), under exclusive license to Springer Nature Singapore Pte Ltd. 2022, corrected publication 2022

D. Bijulal et al. (eds.), *Technology-Enabled Work-System Design*, Design Science and Innovation, https://doi.org/10.1007/978-981-16-1884-0_17

the usage of costly equipment to estimate the body's angular deviation from a neutral position (Kee and Karwowski 2001).

From the literature review, it has been found that awkward postures, high repetition, and forceful exertion are the most impacting risk factors in industrial tasks. Extensive research has been done to assess workplace interventions in order to reduce worker risks of developing work-related musculoskeletal disorders (WMSDs) (Mathiassen and Paquet 2010). But studies evaluating occupational risk factors involved in local and unorganized sectors are limited. In the present research, work-related tasks in the hardware and lock industry have been identified to study, as the majority of workers are employed by these local industries in Aligarh city. An industrial survey was done for the 'on-site' evaluation of work-related physical activities involved in local industries. In the first session, video recordings for workstations and workers' interviews were recorded for seven different substations (electroplating, sieving, punching, grinding, power drilling, polishing, and buffing) of a local hardware industry during their occupational tasks. In the second session, video recordings and interviews were analyzed to determine the task in which musculoskeletal load on workers was the highest. As a result, it has been surfaced that workers in the buffing production line are more prone to develop work-related musculoskeletal disorders as they were sitting under the most awkward posture for a prolonged period. Therefore, in the present study risk factors associated with the buffing task have been analyzed and ergonomic interventions to the workplace are suggested as an attempt to reduce risk factors and subsequently reduce the incidence and magnitude of WMSDs.

2 Method

After a detailed study of the industrial survey, workers interviews, and video recordings, postures for the analysis were considered based on (1) the most awkward posture and difficult work tasks (based on workers' interview and initial observations), (2) the posture sustained by workers for the majority time period, and (3) the posture where the highest force/load occurs (subjective basis). For the study, 16 workers (consideration based on maximum available workers on buffing station in the particular surveyed industry in day shift) doing buffing in the production line were asked to perform the task as they normally do. Then discomfort and postural assessment were carried out. Later, the workstation has been designed in AUTOCAD 2014 and imported in HUMANCAD 2.5 for postural analysis and to validate the subjected and observational results, obtained from discomfort score and REBA score, respectively.

2.1 Discomfort Assessment

Corlett and Bishop's discomfort map (Corlett and Bishop 1976) was used to map different areas of pain in the body and its intensity along with other factors like duration and frequency. The worker performing buffing is shown a body map to indicate the area of pain/discomfort due to the task, and they are free to choose the portion of discomfort on an individual basis.

2.2 Posture Assessment

Among the various available schemes having their benefits and drawbacks, REBA has been used as a postural analysis tool because it can evaluate tasks where postures are dynamic, static or where gross changes in a position take place as in buffing, where workers sit in a particular posture and perform repetitive exertion to buffed the item. As a sensitive tool, it divides the body into parts (neck, upper arm, lower arm, wrist, trunk, and legs). While workers were performing the task, the angle between body parts (neck, upper arm, lower arm, wrist, trunk, and legs) in a neutral position and body parts during the task were measured using the 'Manual goniometer'.

2.3 Existing Workstation Design in AUTOCAD

For analysis purposes, the existing buffing workstation (Fig. 1) has been replicated in AUTOCAD MECHANICAL 2014. Generally, in local industries buffing machines are fitted to the ground and workers are required to sit on the ground while performing the task. But the surveyed industry is a new setup and has a machine base lifted by 340 mm from the ground to a cemented foundation. Similarly, the level of workers' sitting arrangement is also uplifted by providing strong metal benches (not fixed to the ground), but heavy, which prevent its motion against man–machine force and vibrations. The height of benches is 440 mm from the ground, width is 630 mm (Fig. 2),

Fig. 1 Buffing machine

Fig. 2 Bench width

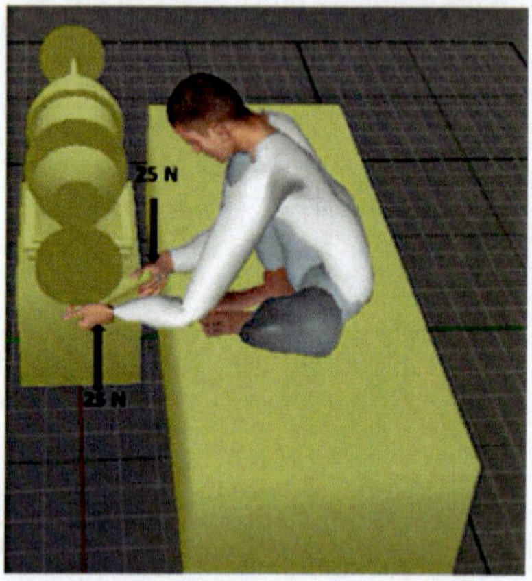

Fig. 3 Posture 1 (worker performing buffing with one leg folded while other leg resting on the station)

and length is about to cover 2–3 buffing stations in continuation. The designed workstation includes a buffing machine and cemented foundation on which it is mounted and sitting space for workers is shown in Fig. 3. The model of buffing workstation is the replica of the actual scenario and meets all the dimensions of the present system.

2.4 Posture Analysis in Human CAD 2.5

The AutoCAD model was imported in HUMAN CAD 2.5 for postural analysis. A human mannequin with Indian anthropometric data was inserted and set in the same posture as workers sit while buffing. A simulated condition was created matching the existing actual environment shown in Figs. 3 and 4. The present HCAD analysis has been done on two major postures which were acquired by most of the workers. Posture 1 is shown in Fig. 3 and posture 2 in Fig. 4. According to the specification of buffing machine (HP-2, RPM-2800, Volts-440, Cycles-50, Phase-3, AMPS-3, Model-232,